DATE DUE

APR 2 3 1996	
MAY 0 6 1997	
JAN - 3 2000	

GAYLORD

PRINTED IN U.S.A.

Introduction to comets

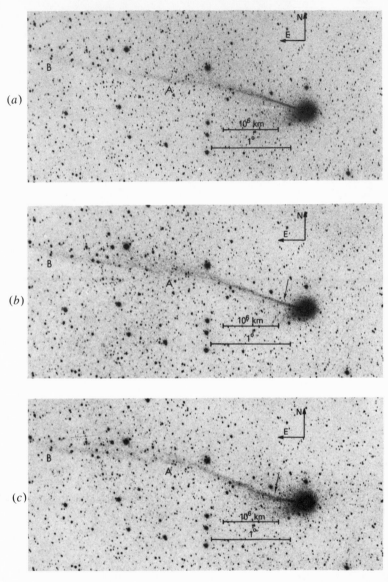

Observations of Comet Bradfield in early 1980, showing a very rapid turning of the plasma tail on February 6. The mid-exposure times are: (a) 2ʰ32ᵐ30ˢ; (b) 2ʰ48ᵐ00ˢ; (c) 3ʰ00ᵐ00ˢ. The change in the position angle of the tail (marked by arrow) was an exceptionally high 10° in 27.5ᵐ. The orientation of the outer tail segment (marked A–B) remains nearly constant. Note that the turning part of the tail lengthens during this sequence; the rate of lengthening corresponds to a speed of approximately 300 km sec⁻¹, a value close to expected solar wind speeds. This event can be simply interpreted by thinking of the plasma tail as a wind sock that shows the direction of the solar wind flow. If so, it was produced by a 50 km sec⁻¹ shift in flow perpendicular to the ecliptic plane from about 30 km sec⁻¹ northward to about 20 km sec⁻¹ southward. (Joint Observatory for Cometary Research, NASA/Goddard Space Flight Center, and New Mexico Institute of Mining and Technology)

Introduction to
Comets

John C. Brandt and Robert D. Chapman

NASA/Goddard Space Flight Center
Greenbelt, Maryland

Cambridge University Press

Cambridge
London   New York   New Rochelle
Melbourne   Sydney

Published by the Press Syndicate of the University of Cambridge
The Pitt Building, Trumpington Street, Cambridge CB2 IRP
32 East 57th Street, New York, NY 10022, USA
296 Beaconsfield Parade, Middle Park, Melbourne 3206, Australia

First published 1981

Printed in the United States of America

Library of Congress Cataloging in Publication Data
Brandt, John C
Introduction to comets.
Bibliography: p.
Includes index.
1. Comets.   I. Chapman, Robert DeWitt, 1937–
joint author.   II. Title.
QB721.B8   523.6   76-47207
ISBN 0 521 23906 0

Contents

Preface

Although comets have been a source of fascination through the ages, they are not presently topics of widespread interest in the astronomical research community. This situation may change, however, in the next decade for several reasons. Interest in comets has been steadily increasing since the 1970s, in part because of observations made from above the atmosphere and also because of our increasing knowledge of the solar wind. Two more or less predictable events in the 1980s should accelerate this trend. The first is the return of the most famous comet of all time, Halley's comet, in 1985 and 1986. The second is the good possibility that a spacecraft will be launched to a comet to perform in situ measurements, and such a mission could involve a landing on a cometary nucleus. The combination of these events should greatly increase the effort in observational and theoretical cometary research.

The purpose of this monograph is to provide an introduction to comets – history, facts, recent results, and future possibilities – for students, scientists, and interested members of the public.

These goals present several practical problems that have been tackled by adopting an unorthodox approach to the organization of the material. An understanding of the historical development of cometary ideas is most useful, and Part I is devoted to an extensive historical review. Knowledge of up-to-date physical models and facts about comets is important, and Part II presents our established picture of comets in an essentially encyclopedic fashion. Hopefully, results derived from new studies of comets will not radically change Part II. An appreciation of recent findings and trends is also of value; these topics are presented in Part III. Here the organization is largely by specific comet, and the topics covered could be expected to change with the appearance of some new comets. Finally, comets are a tremendous source of public interest, and this subject is discussed in Part IV.

References are listed in the Suggested Readings at the end of the text. The Suggested Readings are arranged by topic, following the organization pattern of the text. For convenience, the numerous references for Part II are more finely subdivided. Citations relevant to more than one topic are usually repeated as necessary under the appropriate headings.

Thus, the attempt has been made to produce a monograph focused on the broad view of cometary physics and its interrelationships. Hopefully, this text will be useful as we enter the anticipated Golden Decade of cometary research.

We are indebted to many colleagues both for figures and for helpful criticism. The list is so extensive that we cannot thank everyone personally on these pages. However, three individuals have provided exceptional assistance, and we are pleased to single out for special thanks Armand Delsemme, Donald Yeomans, and Malcolm Niedner.

J. C. B.
R. D. C.

Part I

The historical perspective

On the average, about one naked-eye comet appears every year. Often to observe a comet one has to arise before dawn. However, the effort is well rewarded, for a bright comet is a spectacular sight. There is more to comets than just an enjoyable view. Comets are enigmatic and fascinating celestial objects that might carry clues to the very origin of our solar system. Interest in comets developed early in the history of civilization. It is probable that the earliest interests were more astrological than astronomical. Comets were frequently seen as the indication of coming great events.

The objective of Part I is to trace the historical development of cometary thought from its earliest beginnings through the mid-twentieth century. We will show that studies of comets have gone hand in hand with many of the most significant periods of astronomical thought. In the process of telling the story, we will also develop the modern picture of a comet as a unique celestial object. This part will set the stage for the detailed discussion of cometary physics in the remainder of the text.

To begin the story, try to imagine yourself in the world of 25,000 B.C. You are an unsophisticated wanderer, relying on your skills as a hunter and gatherer to eke out a bare existence. You have become intimately familiar with the sky as a means to tell time and the passing of the seasons. One night, as the sun sets, you see a spectacularly bright comet in the western sky with its spearlike tail pointing away from the horizon. What would be your reaction? Awe? Fear?

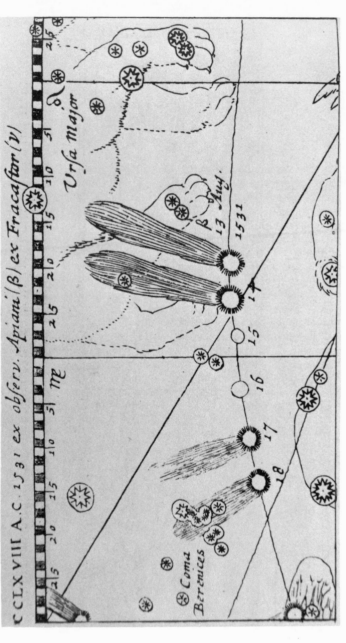

The 1531 apparition of Halley's comet as recorded by Peter Apian. (From Bulletin de la Societé Astronomique de France, 1910)

1   *Comets in history*

Antiquity to the fifteenth century

Prehistoric man must have had a sophisticated knowledge of the night sky, judging from the recent discoveries by archeoastronomers. There can be little doubt that early man came to know the sky – its diurnal risings and settings, its seasonally shifting patterns of stars, its changing lunar phases – as both a clock and a calendar. One can try to imagine the reaction of a prehistoric people to the mysterious appearance of a bright comet in the sky. Probably other notable events would have occurred at the same time – the death of a loved one, the birth of a child, a killing drought, or an especially successful hunt. If they were at all superstitious, early people would have viewed the comet as an omen for whatever important event occurred while it was visible.

When we study the history of man's views of nature, we find that comets have always been surrounded by an aura of awe and mystery. Many people shared Aristotle's view that the appearance of a comet signaled disaster or drought. The appearance of a bright comet struck fear in the hearts of its viewers, and with the fear came considerable interest. Even today, the appearance (or potential appearance) of a bright comet sparks immense public interest; the story of Comet Kohoutek is a case in point.

Over two thousand years ago, the Roman sage Seneca speculated: "Some day there will arise a man who will demonstrate in what regions of the heavens comets take their way; why they journey so far apart from the other planets; what their size, their nature" (Hellman, 1944:33). So far this man has not arisen, and there is little chance that he will arise in the near future. Today, we still do not know the answers to these perplexing questions. Many professional comet workers (the authors included) regard these mysteries as one of the great charms of comets.

Many concepts of the nature of comets were extant during Greek and Roman times. The Greeks gave us the word *comet* (κομήτης), which means "long-haired one." The Latin word for hair (*coma*) has survived as a part of cometary terminology. When one sees a bright comet with its long, wispy tail, it is not difficult to see the origin of the concept of a comet as hairy.

A study of the views about comets held by ancient thinkers is hampered by one serious drawback: We do not often have access to the

3

original writings but must rely on secondhand sources such as Aristotle and Seneca. Even so, we do have some very old records of comets. Among the oldest is a Babylonian inscription interpreted as a reference to the comet of 1140 B.C. "A comet arose whose body was bright like the day, while from its luminous body a tail extended, like the sting of a scorpion."

The history of cometary thought began entirely as a debate over whether comets were celestial objects or phenomena of the atmosphere. The debate started early. The Babylonians (or Chaldeans) are credited with the ideas that comets were cosmic bodies, like planets, with orbits, and that comets were fires produced by violently rotating air. The Pythagoreans (sixth century B.C.) and Hippocrates (c. 440 B.C.) are reported to have considered comets as planets that appeared infrequently and, like Mercury, did not rise very far above the horizon. Hippocrates and his student Aeschylus also believed that the tail was not an integral part of the comet but rather an illusion caused by reflection. Anaxagoras (499–428 B.C.) and Democritus (c. 420 B.C.) thought that comets were conjunctions of planets or wandering stars; Democritus apparently believed that certain stars were left behind when comets dissolved. Ephorus of Cyme (405–330 B.C.) reported that the comet of 371 B.C. split into two stars. Seneca considered this an impossibility and accused Ephorus of spicing up his tales for public consumption. We know today that the splitting of a comet is quite possible (e.g., Biela's comet, Chapter 4). It is not difficult to see how such an observation could lead to the view that comets were formed by a coalescence of stars.

A contemporary of Aristotle, Apollonius of Myndus, rejected the view that comets were an illusion or fire and asserted that they were distinctively heavenly bodies with orbits. Some of the early concepts of comets seem quite reasonable to us today. Certainly, many thinkers viewed comets as celestial objects. However, the influential ideas of almost two millennia were those of Aristotle (384–322 B.C.), as set forth in his *Meteorology* (1952). In this famous work, he first discussed the views of others, then presented his own concepts. Aristotle ruled out the planetary nature of comets by asserting that they had been seen outside the zodiac. In addition, comets could not be caused by a conjunction of planets or a coalescence of stars because many comets had been observed to fade away without leaving behind one or more stars.

Aristotle apparently was impressed with the irregular and unpredictable nature of comets, particularly when contrasted to his philosophical concept of the unchanging nature of the heavens. Hence, he considered that they could not be astronomical bodies but were the product of meteorological processes in our atmosphere; specifically, they lay below the moon. He wrote:

We know that the dry and warm exhalation is the outermost part of the terrestrial world which falls below the circular motion. It, and a great part of the air that is continuous with it below, is carried around the earth by the motion of the circular revolution [the same motion that carries the celestial sphere around the earth]. In the course of this motion it often ignites wherever it may happen to be of the right consistence . . . We may say, then, that a comet is formed when the upper motion introduces into a gathering of this kind a fiery principle not of such excessive strength as to burn up much of the material quickly, nor so weak as soon to be extinguished, but stronger and capable of burning up much material, and when exhalation of the right consistency rises from below and meets it. The kind of comet varies according to the shape which the exhalation happens to take. If it is diffused equally on every side the star is said to be fringed, if it stretches out in one direction it is called bearded. [Aristotle, 1952:450.]

This embryonic classification scheme for comets actually survived at least until books written on the comet of A.D. 1577. Aristotle apparently accepted the idea that comets were omens of droughts and high winds. On Aristotle's own ground, this follows somewhat logically because of the "fiery constitution" of the exhalation. Finally, Aristotle thought that the Milky Way was composed of the same material as the comets.

It is easy to be impatient with and critical of Aristotle's views, but this is not fair. His hypothesis, considered in light of the physics of the era, was a good attempt to explain the sudden appearance, unusual movements, and highly irregular shapes of comets. Aristotle himself considered his explanation satisfactory if it was free of impossibilities. Our ire should be reserved for those investigators 2000 years later who could do no better.

Aristotle's ideas gradually grew in importance. Posidonius (135–51 B.C.) synthesized Aristotle's and added some of his own. Although he regarded comets as atmospheric phenomena, he stated that there were more comets than are usually observed because some are lost in the glare when near the sun. This idea came from an observation made by Posidonius himself of a comet near the sun becoming visible during a total solar eclipse. The classification of comets was according to their shapes. Views similar to Posidonius' were given by Arrian (second century A.D.) in a monograph on comets.

Seneca (4 B.C.–A.D. 65) had a classification scheme similar to those previously mentioned. Although he also reviewed previous knowledge, his writings on comets (found in his *Questiones naturales*) are very different. To us, they seem like those of a scientist assaying a situation, and they are filled with apparent flashes of insight. For example: "I cannot think a comet is a sudden fire, but I rank it among Nature's permanent creations" (Hellman, 1944:31). We also find:

If it were a wandering star [i.e., a planet], says some one, it would be in the zodiac. Who say I, ever thinks of placing a single bound to the stars? or of cooping up the divine into narrow space? These very stars, which you suppose to be the only ones that move, have, as every one knows, orbits different one from another. Why, then, should there not be some stars that have a separate distinctive orbit far removed from them? [Hellman, 1944:32]

Elsewhere we read:

There are many things whose existence we allow, but whose character we are still in ignorance of . . . why should we be surprised, then, that comets, so rare a sight in the universe, are not embraced under definite laws, or that their return is at long intervals? . . . The day will yet come when the progress of research through long ages will reveal to sight the mysteries of nature that are now concealed . . . The day will yet come when posterity will be amazed that we remained ignorant of things that will to them seem so plain. [Hellman, 1944:33]

Even Seneca was somewhat under the influence of his illustrious predecessor. He classified comets under meteorology, and he discussed weather forecasting from the appearance of comets.

Pliny the Elder discussed comets in his *Natural History,* which appeared about A.D. 77; Seneca is not mentioned as a source. Pliny presented a classification scheme based on appearance (both shape and color) that was used for centuries. However, his discussion of comets included little that was new, and many of his statements were not very specific.

It is curious that comets were not mentioned in Ptolemy's (second century A.D.) *Almagest* and were barely mentioned in his other works, and then only in connection with weather prediction. Ptolemy did argue that events on earth were not inevitably influenced by the stars. Arguments of this nature encouraged the notion (which persisted at least into the sixteenth century) that prayers would help avert the undesirable influences of comets.

Cometary studies did not flourish in the centuries following Ptolemy, and Hellman (1944:9) has noted that the years up to the fifteenth century "were not productive of any new cometary theory." Of course, this does not mean that comets were not observed and recorded; appearances of bright comets such as Halley's were recorded, for example, in A.D. 684 in the *Nuremberg Chronicles* and in 1066 in the Bayeux Tapestry (see Figures 1.1 and 1.2). Men such as Bede (A.D. 673–735), Thomas Aquinas (c. 1225–74), and Roger Bacon (c. 1214–94) wrote about comets. Despite variations in individual writings, the astrological view of comets was strengthened, particularly the belief that comets were evil omens. The scientific data recorded in Europe and the Middle East were often just sufficient to identify appearances of the periodic comets. Cometary observations by the Chinese have not been mentioned here because they had little or no influence for centuries on the main development of cometary knowledge.

Figure 1.1. The A.D. 684 apparition of Halley's comet as recorded in the *Nuremberg Chronicles*. (Yerkes Observatory photograph)

Beginning of the fifteenth century to the supernova of 1572

European civilization was slowly climbing out of the medieval period as the fifteenth century opened with the appearance of two bright comets in 1402. Slow progress toward our modern view of comets also began. Some individuals began to study comets in a systematic manner – gathering facts, probing their nature – rather than exploiting them with the superstitious people. How slowly this change came about can be judged, as noted by Hellman (1944:16), by the fact that Pingré in his famous *Cométographie,* published in the 1780s, "still considered it necessary to refute Aristotle." Aristotelian theory was extended in the 1450s by Matthew of Aquila, who associated comets and earthquakes. He also considered that comets not only signaled evil but could cause evil because, in Hellman's words, "their hot, putrid vapors contaminated the air" (Hellman, 1944:73).

Paolo Toscanelli[1] (1397–1482) observed Halley's comet at its 1456 appearance, as well as several other comets. His positional observations for a number of comets that he observed between 1433 and 1472 were sufficiently accurate to permit the calculation of orbits by later

Figure 1.2. The 1066 apparition of Halley's comet as recorded on the Bayeux Tapestry.

workers. Peurbach (1423–61) also observed the comet of 1456 and by attempting to measure its distance may have been the first to do so.

A spate of activity was associated with the great comet of 1472. The principal contribution was made by the legendary but controversial Johannes Müller (1436–76), who is usually known by his adopted Latin name, Regiomontanus. His work on the comet was divided into sections whose titles were a list of problems to be investigated. These include the diameter of the comet, the position of the comet, the length and thickness of the tail, and the distance to the comet by measuring its parallax. His observations were not sufficiently accurate to permit him to infer a meaningful parallax, and his value of 6° is highly erroneous. Nevertheless, either he or Peurbach was the first to attempt such a measurement. In evaluating Regiomontanus' contribution, we must also bear in mind what he did not do. His writings contain no discussions of the "meaning" of the comet, nor did he issue astrological predictions based on the appearance of the comet.

The concept that comet tails point away from the sun was well known by the mid-sixteenth century. The Italian astronomer Fracastoro (c. 1480–1553) wrote *Homocentrica* in 1538 in an attempt to improve upon the Ptolemaic theory of planetary motions by reverting back to concepts of the early Greek thinkers. The effort was not destined to bear fruit, because Copernicus' great *De revolutionibus* was already written and would be published 6 years later. However, Fracastoro did describe his observations of several comets, and he noted that the tails always pointed away from the sun. Fracastoro was not alone in this important observation. Peter Apian described observations of several comets that appeared in the 1530s in his *Astronomicum Caesareum* (1540) and remarked that the cometary tails pointed away from the sun. In his *Practica* (1532), Apian described the appearance of Halley's comet at its 1531 apparition, and a woodcut on the title page shows the tail axis extending through the sun. The figure facing p. 1 of the present text shows Apian's remarkable observations of Halley's comet in 1531.

The close proximity in time of the writings of Fracastoro and Apian has led to some debate over who deserves credit for the discovery. In fact, the exact origin of our understanding of the orientation of comet tails is probably lost in antiquity. Chinese astronomers, in describing their observations of a comet in A.D. 837, said as much, and Seneca in his *Questiones naturales* wrote that "the tails of comets fly from the sun's rays" (Seneca, 1910).

In 1550, Jerome Cardan published his *De Subtilitate*, which was a compendium of learning. Cardan's views on cometary science were of

considerable interest. He was aware of the parallax method for determining the distance to comets. His views on the origin of comets as
summarized by Hellman (1944:93) read: "A comet is a globe formed in
the sky and illuminated by the sun, the rays of which, shining through
the comet, give the appearance of a beard or tail." Cardan also thought
that there were more comets than the ones observed.

The comet of 1533 was observed by Copernicus; unfortunately, the
observations have been lost. We may safely conclude that he made no
significant contribution to cometary knowledge.

A sure sign of at least some maturity of an astronomical subject
appeared at the time of the comet of 1556, when some of the earliest
catalogues of comets were published. At least three such catalogues
were issued within a span of only a few years.

The last key event before the studies of the comet of 1577 was the
supernova of 1572, which was visible for a year in the constellation
Cassiopeia. Hellman (1944:111) explicitly states that "the influence of
the new star of 1572 in moulding the astronomical thought of the period
cannot be overestimated." The principal observer of the supernova
was Tycho Brahe (1546–1601), generally considered to be the greatest
astronomer of his day. Tycho made and recorded positional observations with the high accuracy for which he is renowned. He failed to
detect any parallax, and he placed the new star in the region of the fixed
stars. In retrospect, it is curious that Tycho concluded that the new star
could not be a comet or meteor because these were formed below the
moon.

Tycho's observations were rivaled only by the work of Thomas
Digges (c. 1564–95) in England. Observations were also made by
Michael Maestlin and by Hagecius in Bohemia. The conclusion
reached by all the authors mentioned was that the new star had no
measurable parallax and belonged to the region of the fixed stars. Although there were some dissenting voices concerning the parallax, the
basic result was established, as well as an eager and receptive climate
for the appearance of the comet of 1577.

The comet of 1577

Tycho Brahe observed this famous comet from November 13,
1577, to January 26, 1578. His observations and views were contained
in a Latin work, *De Mundi Aetherei Recentioribus Phaenomenis,* and a
German work. Tycho summarized earlier work on comets and ideas
concerning the structure of the universe. Aristotle's views on the im-
mutability of the heavens were questioned on the basis of the new star
of 1572. Tycho also questioned the atmospheric origin of comets be-

cause, if that concept were true, he could see no reason for their tails always to point away from the sun. Nevertheless, Tycho felt that the conclusive method for discovering the comet's true place was to measure its parallax.

Tycho measured the position of the comet among the stars when the comet was high and low in the sky and compared the results. If the comet was between the earth and the moon, a parallax of at least 1° should have been found. The average error of Tycho's positions (established by comparison with a modern orbit for the comet) was only 4'; we also know that he was well aware of the effects of refraction. Tycho was cautious and concluded that the comet's parallax was 15' or less, placing it at least 230 earth radii away. This distance is well beyond the moon, which averages 60 earth radii away. Comparison of Tycho's observations from the island of Hveen (near Copenhagen) with observations made by Hagecius at Prague showed a difference in position of only one or two minutes of arc. If the parallax was 2', and noting that Prague and Copenhagen are about 600 km apart, the comet would be at a distance of approximately 1 million km, well beyond the moon's mean distance of about 380,000 km.

Thus, Tycho concluded that the comet lay between the moon and Venus. He even attempted to calculate an orbit[2] for the comet. His result was a circular orbit around the sun outside of the orbit of Venus. Of course, he could not represent the observed motion of the comet, assuming it traveled in a circular orbit with uniform motion. He was obliged, therefore, either to assume an irregular motion or to admit that the orbit was not "exactly circular but somewhat oblong, like the figure commonly called oval" (Dreyer, 1953:366). Sarton interprets Tycho as having suggested that the comet's orbit was elliptical. Whether an ellipse or just an oval was meant, Dreyer (1953:366) notes: "This is certainly the first time that an astronomer suggested that a celestial body might move in an orbit differing from a circle, without distinctly saying that the curve was the resultant of several circular motions." Tycho's work *De Mundi Aetherei Recentioribus Phaenomenis* was first published in 1588, and Kepler's result that planteary orbits were ellipses was contained in his book on Mars, which appeared in 1609. The question of the shape of cometary orbits will become more curious when we look at Kepler's own views below.

Tycho attempted to calculate the linear dimensions of the comet, which, of course, depended on the distance assumed. The measured angular dimensions on November 13 were a head diameter of 8', a tail length of 22°, and a tail breadth of 2.5°. Tycho realized that the linear dimensions of the comet were tremendous even if the comet looked small to terrestrial observers.

Tycho expressed his opinion that the tail is caused by sunshine passing through the comet: Because comets are not diaphanous, sunlight cannot pass through without effect and because comets are not thick and opaque like the moon, sunlight is not simply reflected. A comet is intermediate and partly holds the sunshine. Because a comet's body is porous, some sunbeams are allowed to pass through and are seen by us as a tail attached to the head.

Thus, comets were not atmospheric and were not sublunar. They were supralunar celestial objects that could be studied by scientific methods. The lack of appreciable parallax clinched these conclusions. There was really no need to take the Aristotelian view seriously thereafter. Tycho's work was confirmed by astronomer Michael Maestlin (1550–1631) and by others.

Of course, there were dissenters from Tycho's conclusions about the comet of 1577. Of first importance was Tadeáš Hájek Hájku, known by his latinized name of Hagecius (c. 1526–1600); he was considered a leading astronomer of his time. He initially obtained a parallax of 5°, which would have placed the comet well below the moon and, in fact, only about 8 radii from the earth's center. Hagecius' observations were good, and Tycho used them to establish independently the supralunar position of the comet. Hagecius had erred in his interpretation of his imprecise observations. He established himself as a man of considerable scientific character by admitting that he was wrong. He recognized the comet as supralunar in his work on the comet of 1580.

Other observers found a large parallax for the comet, but the quality of this contemporary work was not close to Tycho's. The most serious attacks on Tycho would occur after his death.

Beginning of the seventeenth century to the comets of 1680 and 1682

Johannes Kepler (1571–1630) worked on the comets of 1607 (Halley's) and 1618, and published his views on comets in *De Cometis* (1619) and in *Hyperaspistes* (1625). His ideas on the formation of comet tails and the ultimate extinction of comets seem remarkably modern. He wrote:

Gross matter collects under a spherical form; it receives and reflects the light of the sun and is set in motion like a star. The direct rays of the sun strike upon it, penetrate its substance, draw away with them a portion of this matter, and issue thence to form the track of light we call the tail of the comet. This action of the solar rays attenuates the particles which compose the body of the comet. It drives them away; it dissipates them. In this manner the comet is consumed by breathing out, so to speak, its own tail. [Olivier, 1930:9]

Today we believe that comets shine by reflected light and that their tails are formed by solar radiation pressure. Kepler also considered that comets were as numerous as fish in the sea.

Kepler's beliefs concerning the orbits of comets are curious indeed. He thought that comets moved along straight lines, but with an irregular speed. Because Tycho had suggested an oval or elliptical orbit for comets, and because Kepler himself found elliptical orbits for the planets, this oversight is all the more puzzling.

Kepler would once again become involved in comet work because of attacks on Tycho by Scipio Claramontius (1565–1652) (sometimes called Chiaramonti) and by Galileo. In Claramontius work, *Antitycho* (1621), he attempted to prove that comets were sublunar. Among other things, Claramontius may not have been straightforward because, as Drake and O'Malley (1960:374) have noted: "One of the favorite devices of Chiaramonti was to attack the data of astronomers as inaccurate because they were never in exact agreement. On this pretext, he would throw out all the observations which did not suit his purpose." Claramontius convinced only a few, and his writings did not greatly hinder acceptance of the new approach to comets.

The criticisms of Galileo Galilei (1564–1642) were more serious. Three comets appeared in the year 1618 and an exchange of differing opinions began; the exchange produced Galileo's work dealing with comets, *The Assayer* (1623). Galileo was not in good health, and he did not personally observe the comets extensively. Perhaps he used comets to publicize his scientific methods. In any case, Galileo's antagonism toward Tycho shows clearly in *The Assayer*. Galileo dismissed the parallax observations by noting that the comet could be simply an optical illusion. He suggested that vapors rising from the earth's atmosphere could produce comets seen by reflected sunlight when the vapors had risen outside the cone of the earth's shadow. A clever argument by Father Horatio Grassi (1583–1654) was that comets were quite distant because they were not magnified by telescopes; Galileo argued that this was not necessarily so if strange optical effects were involved. The curvature of some comet tails caused great difficulties to all concerned.

All in all, the writings concerning the comets of 1618 are not enlightening. For example, luminous gas was supposed to be both opaque and transparent by the opposing sides. It is painful to read that the biblical description of Shadrack, Meshack, and Abednego walking uninjured in the midst of fire after they had been cast into the furnace was used as evidence by both sides.

Kepler responded to the unjust criticisms of Tycho by Claramontius in his book *Tychonis Brahei Dani Hyperaspistes* or *The Shieldbearer to Tycho Brahe the Dane* (1625). He disposed of Claramontius' views on

Figure 1.3. Three allegorical figures showing the Aristotelian idea that comets are sublunar (left), the Keplerian notion that comets move in a straight line (right), and the idea of Hevelius (center) that comets originate in the atmospheres of Jupiter or Saturn and move about the sun on a curved trajectory. (Frontispiece from Hevelius's *Cometographia*, published 1668 in Danzig.)

the parallax by noting the well-established accuracy of Tycho's observations. Kepler responded to Galileo in an appendix to the *Hyperaspistes*. This reply contains some of Kepler's ideas on comets, mentioned above.

Newton, Halley, and the orbits of comets

The problem of cometary orbits began to be resolved in 1665 when Giovanni Borelli (1608–79), using a pseudonym, published the suggestion that the paths of comets were parabolic. The same suggestion was made by Johannes Hevelius[3] (1611–87), who discovered four comets and wrote two books on comets (Figure 1.3). His ideas on the motion of comets involved an origin in the atmospheres of Jupiter or Saturn and a resisting medium in interplanetary space. A comet began its orbit toward the sun with the flat part of its "disk" oriented perpendicularly to the direction of motion. But when the comet approached the sun, it moved with the edge of the disk forward. The reduced effect of the resisting medium caused a departure from the initial rectilinear path; the resulting path could be either a parabola or a hyperbola, although Hevelius did not put the sun at the focus. The comet of 1680 was studied by a student of Hevelius, George Dörffel, who suggested that its orbit could be represented by a parabola with the sun at the focus.

Isaac Newton (1642–1727) wrote the great book *Philosophiae Naturalis Principia Mathematica (Mathematical Principles of Natural Philosophy)* or *Principia,* which was published in several editions, the first in 1687, the second in 1713, and the third in 1726. The *Principia* was translated from Latin into English by Andrew Motte in 1729. This English rendition of the third edition was revised and supplied with a historical and explanatory appendix by Florian Cajori and published together with Newton's *System of the World* in 1934. Comets are an important part of both these books. Edmund Halley, who was instrumental in seeing that the *Principia* was published, played an important role in Newton's interest in comets. Halley's contribution was diplomatic and financial as well as scientific, moving A. De Morgan (1806–71) to write: "but for him, in all human probability, that work would not have been thought of, nor when thought of written, nor when written printed."

Newton's laws of motion and gravitation established the basis for calculating cometary (as well as planetary) motion. Newton gave a method for computing parabolic orbits for comets from three observations. He knew that most comet orbits near the sun could be represented very accurately by parabolas. In 1695, Halley began to calcu-

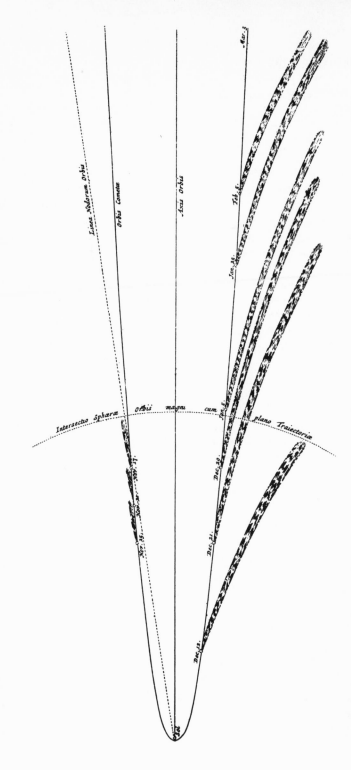

Figure 1.4. Halley's orbit for the comet of 1680 as given in Newton's *Principia*.

late the orbits of several well-observed comets. The orbit for the comet of 1680 was discussed extensively in the *Principia,* and a diagram representing the comet's orbit and tail orientation was presented (Figure 1.4).

In referring to this orbit and the agreement with observation, Newton wrote in the *Principia:* "The orbit is determined . . . by the computation of Dr. Halley, in an ellipse. And it is shown that . . . the comet took its course through the nine signs of the heavens, with as much accuracy as the planets move in the elliptic orbits given in astronomy" (Newton, 1686:Book III). Thus, Newton concluded that "comets are a sort of planet revolved in very eccentric orbits around the sun."

Newton's views on the physical constitution of comets are also very interesting. He held that comets shine by reflected sunlight, a fact that also explained why comets were usually observed near the sun. Newton reviewed the historical ideas concerning comet tails and concluded that they arose from the atmospheres of the comets. He felt that this would not be difficult because "a very small amount of vapor may be sufficient to explain all the phenomena of the tails of comets" (Newton, 1686:606). Newton also suggested that a nova could be produced by the infall of combustible cometary material onto a star.

Newton's views on the orbits of comets were not immediately accepted by everyone. All reasonable doubt would be dispelled by future observations of the comet of 1682. Halley (Figure 1.5) found the orbits of about two dozen comets for which there were sufficient observations and published a catalogue of their elements in 1705. The orbital elements (Chapter 3) for the comet of 1682 showed close correspondence with the comets of 1607 (observed by Kepler and Longomontanus) and the comet of 1531 (observed by Apian). Halley also knew that the great comet observed in the summer of 1456 traveled in a retrograde direction. Although the periods involved in these identifications showed variations (roughly from 75 to 76 years), Halley concluded that all the observations referred to the same comet and wrote: "I may, therefore, with confidence, predict its return in the year 1758. If this prediction is fulfilled, there is no reason to doubt that the other comets will return" (Armitage, 1966:166). He also wrote that if he were correct, "candid posterity will not refuse to acknowledge that this was first discovered by an Englishman" (Armitage, 1966:166).

Halley knew that the planets Jupiter and Saturn would disturb the orbit of his comet. This was expected from Newton's law of gravitation and was the physical reason for the differences in the revolution period. However, detailed perturbation calculations were not made in England but in France, by the astronomer A. Clairaut (1713–65). The job was herculean because all computing had to be done by hand. Clairaut and his associates completed their calculations and reported their results in November 1758. The date of perihelion passage was calculated to be April 15, 1759, with an estimated uncertainty of 1 month.[4] The comet was observed on Christmas night 1758 by an amateur astronomer living near Dresden. It passed perihelion on March 13, 1759.

At its 1759 apparition Halley's comet was observed by numerous astronomers, including Charles Messier (1730–1817). But none of the observations at that time compare in importance with the fulfillment of

Figure 1.5. Edmund Halley. (Yerkes Observatory photograph)

Halley's prediction, with its vast philosophical implications for astronomy in general and comets in particular. Comets were shown to be subject to the laws of physics. Their orbits could be calculated and their return predicted years in advance. At least some comets were members of the solar system.[5] Any rational fear of comets as signs of disaster or evil should have vanished.

The return of Halley's comet as predicted was also a triumph for the detailed physics whose development was initiated by Newton and is now called *celestial mechanics*. In other words, Newton's physics[6] clearly applied out to the aphelion distance of Halley's comet, about 35 AU.

The understanding of cometary orbits was a strong argument against René Descartes's (1596–1650) vortices, which were used to explain planetary motion. It is difficult to see how a vortex could pull planets along their orbits, and yet retrograde comets such as Halley's could move in the opposite direction to the planets without any discernible effect.

We should also note that any residual support for the concept of solid spheres as the means of carrying the planets around the sky in the Ptolemaic system was shattered by knowledge of the high ellipticity of cometary orbits. The first edition (1769) of the *Encyclopaedia Britannica* states that comets "have moved through the etherial regions and the orbits of the planets without suffering the least sensible resistance in their motions; which plainly proves that the planets do not move in solid orbs."

Halley's data on the orbits of comets showed clearly the contrast between cometary and planetary orbits. The elliptical orbits of comets were very elongated (such that parabolas were a good approximation near the sun), whereas the elliptical orbits of planets showed only nominal elongation (such that the circles were a fairly good approximation of their orbits). Whereas the planes of planetary orbits were closely confined to the plane of the earth's orbit and all went in the same (direct) direction, the planes of cometary orbits could be inclined at all angles to the ecliptic, and their direction of motion could be either direct or retrograde.

The development of the orbit calculation phase of celestial mechanics proceeded relatively rapidly. P. S. Laplace (1749–1827) gave formulas for the calculation of elliptical orbits through the method of successive approximations. Although elegant and rigorously correct, the method was unwieldy and not satisfactory. The practical problem was solved in 1797 by the physician and respected amateur astronomer Wilhelm Olbers[7] (1758–1840). His simple method for determining the five elements needed for a parabolic comet orbit has not been significantly improved to the present day.

The second person to successfully predict the return of a comet was Johann Encke (1791–1865). He studied the available orbital information for comets seen in January 1786 by Mechain, in November 1795 by Caroline Herschel, in October 1805 by J. Pons and others, and by Pons again in November 1818. No parabolic orbit would fit the observations of the comet observed in 1818, and this situation stimulated Encke to undertake a thorough study of the observations. He soon realized that all the observations mentioned above referred to the same comet, which moved in a quite unusual elliptical orbit. The comet's period of revolution was 3.3 years, and its heliocentric distance varied between 0.34 and 4.08 AU. Encke computed the orbit using the method developed by Carl Friedrich Gauss (1777–1865). The method was published in Gauss's *Theoria motus corporum coelestium* (1809). It permitted calculation of all six orbital elements from three observations of position. Gauss had developed his method as an aid in the search for the so-called lost planet Ceres.

The comet studied by Encke was predicted to return in 1822, and it was observed from Australia. *Encke's comet,* as it is now called, has been repeatedly observed to this day.[8]

The 1838 return of Encke's comet presented the astronomical community with another problem. As Encke had suspected for some time, the comet's period was steadily decreasing at an accelerating rate. On successive returns the comet would arrive at perihelion earlier than predicted by an amount that varied but that was typically about 0.1 day. There were three ways of explaining the observations. (1) The comet could be moving through an essentially uniform resisting medium (proposed by Encke). In such a situation, the comet would spiral in toward the sun while the period decreased. (2) A belt of meteoric particles[9] orbiting the sun would serve the same purpose as Encke's resisting medium. However, variable decreases would occur because comets would be affected only while passing through the belt. In both of these theories, the period of the comet could only decrease. (3) The deviations could be due to the expulsion of material from the comet in a particular direction, that is, a "rocket effect." This was suggested by Bessel after observing Halley's comet in 1835 and noticing a sunward plume of material that resembled a blazing rocket. Bessel actually computed the orbital consequences of a simple model for a sunward rocket.

Subsequent work would show that the hypotheses of the resisting medium and the meteoric belt both had to be abandoned because comets were found that had increasing periods. Bessel's basic idea has survived and has been incorporated in our current physical picture of comets (Chapter 4).

Shortly after the predicted return of Encke's comet in 1822, another comet was discovered in 1826 that would rival it for historical interest. Investigations on the orbit of the comet discovered by M. Biela indicated that a parabola would not fit the observations, but that an elliptical path with a period of 6.75 years would; the comet's aphelion was near but outside the orbit of Jupiter. The comet was predicted (by several astronomers, including Olbers) to return in November 1832, and it appeared on schedule. Detailed calculations of the orbit had to include the strong effects of Jupiter. The results indicated that the comet had been sighted twice previously, by Montaigne (and also by Messier) in 1772 and by Pons in 1805. The comet's return in 1839 was not observed because the relative geometry always kept the comet so near to the sun that it could not be seen.

The return of Biela's comet in 1846 was proceeding comfortably when, in the words of John Herschel (1792–1871), "it was actually seen to separate itself into two distinct comets, which, after thus parting company, continued to journey along amicably through an arc of upwards of 70° of their apparent orbit, keeping all the while within the same field of view of the telescope pointed towards them" (Herschel, 1871:390).

At first there was some kind of interaction between the two Bielas, including luminous bridges. Eventually, both parts of Biela developed into complete comets with tails.

Both Bielas returned in 1852, separated by approximately 2 million km. They may have returned in 1858, but as in 1839, their orbits were too near the sun to permit observation. The next scheduled appearance for Biela was in 1866, and there was considerable interest in determining the separation of the two parts. Biela's comet was never seen again. Some astronomers expected to see the debris of Biela's comet in 1872 and, in a sense, they were not disappointed.

The tremendous meteor shower (Leonids) on the night of November 12–13, 1833 (Figure 1.6), stimulated scientific interest in meteors. Several observers noted that they all seemed to come from the same point in the sky located in the constellation Leo. It was suggested that the original particles were traveling in parallel paths in space. Hence, there was considerable interest in their orbits.

In 1866, Giovanni Schiaparelli (1835–1910) established the connection between the orbits of meteor streams and the orbits of comets (see Figure 6.1). He showed the link between the orbits of the Perseids and Comet Swift–Tuttle (1862 III) and between the orbits of the Leonids and Comet Tempel–Tuttle (1866 I). The natural supposition[10]

Figure 1.6. Woodcut showing the Leonid meteor shower of 1833. (Courtesy of the American Museum of Natural History)

was that the particles producing the meteor showers were the last remains of the comets. Biela's orbit was identified in 1867 (by Weiss and d'Arrest) with the Andromedid meteor shower, now also called the *Bielids*. A shower was predicted for 1872, and a wonderful display occurred. Another excellent display appeared in 1885. The shower was weaker in 1892 and apparently expired in 1899 (although a few possible stragglers are recorded from time to time).

We have seen that a basic understanding of cometary orbits was achieved in the nineteenth century. Most comets were thought to travel in very elongated ellipses or perhaps parabolas (i.e., the so-called nonperiodic comets), which were oriented at random and were not confined to the ecliptic plane. The short-period comets[11] had been discovered with aphelia near the orbit of Jupiter. These orbits would be compatible with J. L. Lagrange's (1736–1813) suggestion that comets were ejected from Jupiter or with P. S. Laplace's statement that comets could acquire an orbit with a period ≈ 5 years by a close encounter with Jupiter. Further orbital studies would be needed to sort out the various ideas on the origin of comets. The identity of the orbits of some meteor streams with specific cometary orbits was established.

In many respects the determination of the basic facts of cometary orbits was a triumph of Newton's celestial mechanics; predictions such as the return of Halley's comet in 1759 were spectacular. However, total triumph was short-lived, because early in the nineteenth century significant departures from the predictions for some comets were discovered. The problem of the nongravitational forces on comets would not be understood until the middle of the twentieth century.

In this chapter, we have traced the history of cometary science up to the late nineteenth century. In the next chapter, we will trace the historical development of cometary physics from the closing years of the nineteenth century up to the present time. That chapter will serve as a brief introduction to cometary physics, which will be developed in detail in Part II.

2 Development of modern ideas on the physics of comets

In the last chapter, we traced the history of cometary physics up to the closing years of the nineteenth century. The late nineteenth and early twentieth centuries were a period of revolution in physics during which many of our modern conceptions developed. During the same period, our understanding of cometary physics increased rapidly. The modern era of cometary physics, which began in 1950, built upon the increased understanding of physics and the growing body of cometary data. Our objective in Chapter 2 is to present a summary of our knowledge about comets circa 1950. This will provide the groundwork for the detailed discussion to be presented in Part II.

Physics of comets

Forms and tail structure

Although Newton and Olbers had made efforts to understand the physical nature of comets, this phase of cometary research began most seriously with the research of F. W. Bessel (1784–1846). Bessel was an active observer of Halley's comet in 1835. He noted extensive fine structure near the nucleus consisting of jets, rays, fans, cones, and other forms. Similar structure was also observed by F. G. W. Struve (at Dorpat), John Herschel (at the Cape of Good Hope), and M. Arago (at Paris). A cone of light extending toward the sun for a short distance was observed to curl backward away from the sun as if propelled by a great force. Both Bessel and Herschel (as well as Olbers earlier) hypothesized that the force acting on the material in the cone could be electrical, possibly due to a net charge on the sun.

Bessel developed the theory of cometary forms for particles leaving the head of a comet while subjected to different repulsive forces and solar gravity. This approach was followed and extended by F. A. Bredichin (1831–1904). The mechanical treatment of cometary forms is still known as the Bessel–Bredichin theory. Bredichin classified comet tails into three types, depending on the degree of curvature, and assigned each a specific chemical composition, namely, hydrogen to type I, hydrocarbons to type II, and metallic vapors to type III. Type I tails were nearly straight (curvature hard to measure), and type III tails were strongly curved. The force of repulsion was highest for the type I tails and lowest for the type III tails. Of course, many details of Bredichin's classification scheme have not survived. The Bessel–Bredichin theory

24

of tail forms was extended in the first half of the twentieth century by A. Kopff and by S. V. Orlov, who also gave a fairly elaborate classification scheme.

A candidate for the repulsive force was suggested in 1900 by the Swedish chemist Svante Arrhenius (1859–1927). His suggestion was the radiation pressure of sunlight. Calculations made in the first decade of the twentieth century by Karl Schwarzschild (1873–1916) and by Peter Debye (1884–1966) indicated that the force on a spherical dust particle was at maximum when the diameter was approximately one-third the wavelength of the incident radiation, and the force of repulsion could be 20 to 30 times the solar attraction. Debye also pointed out that the radiation pressure on a single molecule due to selective absorption and reemission of photons could be even higher.

The origin of the idea that the solar repulsive force could be caused by particles emitted from the sun is hard to pin down, but the idea was certainly extant around the turn of the century. The impetus for much of the work on solar particle emission came from studies of the aurorae and related geophysical phenomena. The solar physicist Richard Carrington (1826–75) observed the first recorded solar flare on September 1, 1859, and noted an intense aurora the following day. Carrington was fully aware of the possible cause-and-effect relationship, but he was wary of jumping to this conclusion. A. H. Becquerel (1852–1908) thought that sunspots might be the source of the auroral particles. In 1892, G. F. FitzGerald (1851–1901) estimated the speed of the particles at "about 300 miles per second" from the time between the central meridian passage of a sunspot and the magnetic storm associated with it. In a fascinating paper written in 1900 by Oliver Lodge (1851–1940), the aurorae, magnetic storms, and behavior of comet tails were attributed to "a torrent or flying cloud of charged atoms or ions."

Thus, in 1910, A. C. D. Crommelin wrote:

There are at least three theories to explain the repulsion of the tail from the sun: (1) Light-pressure; (2) Electrical repulsion; (3) Mechanical bombardment by electrons, or other tiny particles violently ejected from the sun. It is quite possible that all three act conjointly, as no one of them seems capable of explaining all the facts. [Proctor and Crommelin, 1937:189]

About the same time, some problems and confirmations began to emerge in connection with the simple application of the Bessel–Bredichin (mechanical) theory of cometary forms. The stimulus was cometary photography (see Chapter 4). A. Pannekoek wrote in his *A History of Astronomy:*

comet tails, formerly smooth and ghostly, hardly visible phantoms, now appeared on the plates as brilliant torches with rich detail of structure, with bright and faint spots, never before seen or even suspected. Such photos, taken of

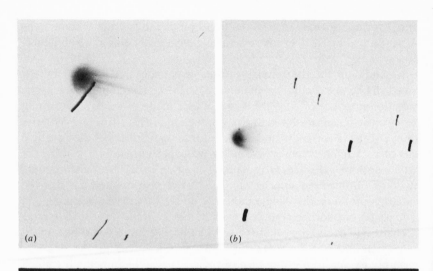

every succeeding bright comet (like Morehouse's comet in 1908 and Halley's in 1910, often reproduced in scientific and popular reviews) gave a new impulse to the study of comet tails. [Pannekoek, 1961:425]

Comet Morehouse (Figure 2.1) showed striking parabolic envelopes on the side toward the sun, which were analyzed by A. S. Eddington (1882–1944). Their formation was relatively easy to understand on the basis of the *fountain model*, in which particles were ejected in various directions toward the sun and then repulsed into the tail. The outer envelope of the particle trajectories formed the observed parabola (Figure 2.2). The problem with the simple interpretation was that it required very large initial ejection velocities and repulsive forces some 800 times solar gravity. Eddington himself was skeptical of these high forces. Also, in Comet Morehouse, E. E. Barnard (1857–1923) noted that specific parts of the tail would rapidly brighten where no material was previously visible and with no obvious supply from the nucleus.

Knots or condensations in the tail of Halley's comet (1910) were studied by Heber D. Curtis (1872–1942), and the same feature could be identified on the photographs from successive nights (Figure 2.3). Curtis calculated the velocities of these knots; the values found range from 5 km sec^{-1} near the nucleus to about 90 km sec^{-1} nearly 0.1 AU from the nucleus. Here the action of the repulsive solar force could be seen directly.[1] These increasing velocities of knots in Halley's comet, as well as those observed in many other comets, could be used to determine the solar repulsive force. The average value of some 200 times solar gravity was considered quite high; there was substantial variation around the average, with some values approaching 1000 times solar gravity. In addition, different clouds observed at the same time in the same comet tail could show quite different repulsions.

Despite the problems with some aspects of the mechanical theory and despite the different suggestions for the repulsive force, the mechanical theory with radiation pressure as the repulsive force would rule the scene for decades. Russell, Dugan, and Stewart in their classic *Astronomy* (1926) stated:

As the activity increases, the finer particles, repelled by the sun's light-pressure, stream away visibly to form the tail, which grows longer and brighter as the comet approaches perihelion. The emissions from the nucleus sometimes take the form of jets, or streams, and sometimes the outflow is more regular, resulting in the formation of envelopes. [Russell, Dugan, and Stewart, 1926:444]

Figure 2.1. Parabolic envelopes in comets. (*a* and *b*) Photographs of the head region of Comet Morehouse (1908c) showing envelope structures. (Greenwich plate, reproduced by K. Wurm, courtesy of J. Rahe); (*c*) Head region of Donati's comet in 1858, drawn by G. P. Bond, showing envelopes. (Yerkes Observatory photograph). See also Figure 2.2.

As cometary spectroscopy developed, we learned that the long, straight tails (type I) were composed not of dust but of ionized molecules. The principal emission in the visual wavelength range comes from CO^+. In 1943 Karl Wurm published his results on the repulsive forces from radiation pressure for CO^+ as well as for C_2 and CN

Figure 2.2. Relationship between particle orbits and cometary structures. (*a*) In the head, the envelope of particle orbits (dashed lines) forms the observed parabolic structure. (*b*) Comparison between particle orbits and tail shape. Particles emitted at times $t_1, t_2, t_3,$ and t_4 are at positions $a_1, a_2, a_3,$ and a_4 at t_5 to form the tail.

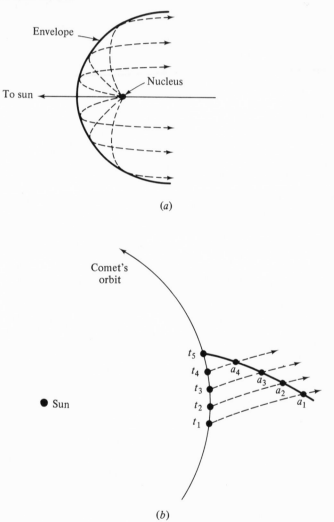

Figure 2.3. Halley's comet in 1910 as photographed at Williams Bay, Wisconsin, June 6, 15.8^h GMT (top); Honolulu, Hawaii, June 6, 18.5^h GMT (center); and Beirut, June 7, 7.0^h GMT (bottom). (Yerkes Observatory photograph)

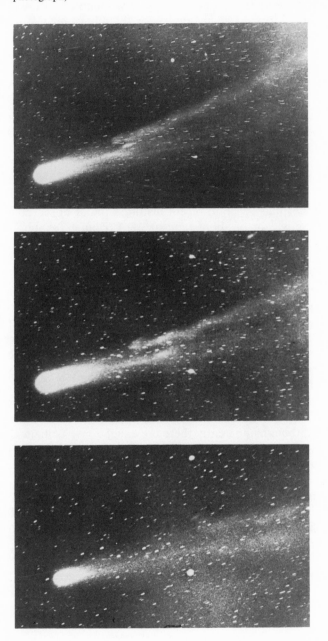

(cyanogen), which are found in the head. This calculation requires knowledge of the solar radiation flux in the relevant wavelength region and the atomic parameters – particularly transition probabilities (f values) – for the molecules involved. Wurm found repulsive forces of 47.0, 1.7, and 0.7 times solar gravity for the molecules CO^+, C_2, and CN, respectively. The low values for C_2 and CN mean that the motions and lifetimes of these molecules are consistent with the observation that they make up the cometary head. The calculated value of 47.0 for CO^+ certainly indicates that this ionized molecule should be driven into the cometary tail, but the value is too low to explain the observed high values of the repulsive force. This problem is not yet completely solved quantitatively, but we will see that it has been approached anew in the modern era (Chapter 4) of cometary research.

Determination of material densities in comet tails depends on spectroscopic and photometric observations. Some idea of the diffuseness of comet tails comes from the fact that the earth has probably passed through comet tails twice in recent history without observable effect. The first time was on or about June 30, 1861 (Comet Tebbutt), and the second time was on or about May 20, 1910 (Halley's comet). It is difficult to be precise about the exact times because the comet tails are not exactly straight and because there are no exceptional meteorological phenomena to mark the passage. In 1910 the earth apparently did not pass through the central part of the main tail, but only through its outer edges or a detached streamer. Astronomically, the passage through the tail produced some spectacular views of the tail stretching across the sky, and little else. Unfortunately, the event caused concern and fear because of the belief that poisonous gases from the comet would contaminate the atmosphere.

The very low densities in comet tails are confirmed by the fact that stars generally can be observed through them without diminution or deflection, although apparently there are exceptions. William Herschel (1738–1822) reported that some diminution occurred for stars seen through the tail of the comet of 1807 and that considerable diminution was found for a star seen through the head. By contrast, no effect was observed by either Struve or Bessel for stars seen quite close to the apparent nucleus of Halley's comet in September 1835. A star field photographed through the head of Comet Pons–Winnecke by Georges Van Biesbroeck (1880–1974) in June 1927 showed no shift in position greater than $0.05''-0.10''$.

It appears that densities in the head and particularly in the tail are insufficient to dim the light from stars measurably or to shift their position. Exceptions occur, but these would be expected primarily when the star is nearly coincident with the apparent nucleus.

Spectroscopy

As the discussion in the last section showed, progress in the physics of comets depended crucially upon the spectroscopic identification of the chemical constituents involved. Work in cometary spectroscopy began in the 1860s.

On August 5, 1864, Giovanni Donati (1828–73) made the first spectroscopic observations of a comet. When he turned his visual spectroscope toward Comet Tempel (1864 II), he saw three faint bands of emission. Sir William Huggins, (1824–1910), one of the pioneering stellar spectroscopists, observed a cometary spectrum in 1866. Not only were the three bands observed by Donati visible, but a reflected solar continuum was also present.

Cometary spectra were intensively observed and investigated by Huggins, and in 1868 he identified the bands observed by Donati. Swan had seen these same bands in hydrocarbon[2] compounds (such as ethylene vapor, C_2H_4) excited by electrical discharges. These bands also could be observed in the gas obtained by heating meteorites, a result that seemed natural if comets were actually a swarm of meteorites. Huggins made the identification by carefully comparing the comet's spectrum directly with the spark spectrum of ethylene vapor; that is, the two spectra were observed side by side in the same eyepiece. Huggins noted:

the apparent identity of the comet's spectrum with that of carbon resides not only in the coincidence of the positions of the bands in the spectra but also in the very remarkable resemblance of the corresponding bands in what concerns their general characters, as well as their relative light. This is well recognized in the middle (green) band where the gradation of intensity is not uniform. [Olivier, 1930:80]

Actually, Huggins needed some good fortune to complement his care. The Swan band spectrum of C_2 shows a similar intensity distribution in both the laboratory sources and comets. This coincidence does not generally hold for all chemical species or physical conditions.

Photographic records of cometary spectra were first made in 1881. Again, Huggins was the pioneer with his spectrum of Comet 1881 III taken on June 24, 1881.[3] The early photographic results showed (in addition to the Swan bands of C_2) a strong group of emission bands in the violet, which were subsequently identified as CN.

The sodium D lines were first detected in the spectrum of Comet 1881 III and shortly thereafter in three other comets with small perihelion distances. The radial velocity of a comet was measured for the first time in 1882 using Doppler shifts of its D lines. Also in 1882 iron lines were observed in the spectrum of a comet that approached very close to the sun.

By the turn of the century, the heads of comets were considered to show the following typical spectra: (1) the (Swan) emission bands of carbon (C_2); (2) the violet bands of cyanogen; (3) emissions near 4050 Å and 4310 Å, which would be identified as C_3 and CH, respectively; (4) the sodium D lines, usually for comets near the sun; and (5) the continuous solar spectrum with its strongest Fraunhofer lines. Little was known about the tail spectrum. Of course, there were exceptions to the typical spectrum. Comet Holmes in 1892 showed a continuous spectrum only, whereas Comet Brorsen in 1868 did not show the Swan bands.

The theory of atomic and molecular spectra was still essentially nonexistent by 1900. An understanding of cometary spectra was, of course, dependent on parallel development in the theory. Even the mechanism of emission was not understood. Were comets self-luminous or did they shine because of the sun's radiation? The basically correct answer was given by K. Schwarzschild and M. Kron in

Figure 2.4. Objective-prism spectrum of Comet Morehouse obtained by Frost and Parkhurst. Note the doublet structure here and compare with modern spectra shown in Figure 4.2. The streaks across the images of the comet are stellar spectra. (Yerkes Observatory photograph)

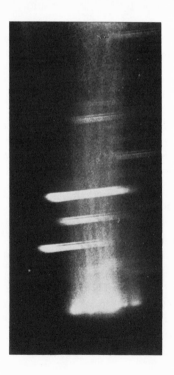

their 1911 paper on the intensity distribution in the tail of Halley's comet.[4] They attributed the mechanism to the absorption and reemission of solar radiation, that is, fluorescence. Although there was some debate over the years, Herman Zanstra (1894–1972) in 1929 showed that fluorescence (and resonance fluorescence) fully accounted for the line and band spectra of comets.

Information on the composition of comet tails began to accumulate early in the twentieth century with the beginning of observations with objective prisms. The leader in this work was Fernand Baldet. Spectra of Comet Daniel (1907 IV) and Comet Morehouse (1908 III) were obtained, and they showed emission extending into the tails. Comet Morehouse had no detectable continuum in the tail, but the emission bands appeared as doublets (Figure 2.4). These doublets were duplicated in the laboratory by A. Fowler and subsequently were shown to be due to ionized carbon monoxide, CO^+. In addition, N_2^+ emission was found in comet tails by the early investigators.

In subsequent years the basic picture would be built up and many new identifications established. The C_2 and CN emissions are confined to the head and are the first (other than the continuous spectrum) to appear as a comet approaches the sun. As the comet nears 1 AU, CO^+ appears in the tail and, if the comet goes near the sun, the sodium lines appear in the head, whereas the Swan bands (C_2), for example, decrease. If the comet passes within 0.1 AU of the sun, lines of iron and nickel appear. As the comet recedes from the sun, the sequence occurs in reverse order (see Figure 2.5).

We conclude this section on the history of spectroscopy by mentioning an important consequence of the fluorescence mechanism. The intensity distribution within the molecular bands found in comets usually did not resemble the distribution observed in laboratory sources. The problem of understanding band structures was compounded by the fact that the intensity distribution was significantly different from comet to comet and even varied within the same comet at different times or at different heliocentric distances. The phenomenon, now called the *Swings effect,* was investigated and understood by Pol Swings in 1941. He pointed out that the exciting radiation for the fluorescence mechanism was the solar continuum, which showed variations with wavelength particularly near strong Fraunhofer lines. In addition, the radial velocity of a comet relative to the sun would cause a Doppler shift of the Fraunhofer lines with respect to the cometary band systems. Thus, a particular line in a cometary band could be bright or faint depending on its position relative to a strong solar absorption line. Swings's hypothesis was confirmed by A. McKellar's observations of Comet Cunningham (1941 I).

Nuclei and cometary phenomena

The need for a cometary nucleus or nuclei was realized very early in the historical game. The dust and gas observed in the coma and tail had to have a source. It became clear very early that the material in the coma and tail was completely lost to a comet and had to be continually replenished. This conclusion was reinforced by observations of Comet Perrine (1897 III) and Comet Ensor (1926 III), both of which faded and vanished as they moved in the inner solar system. The

Figure 2.5. Spectra of Comet Cunningham (1940c) at different heliocentric distances (*r*). The line labeled *NG* (nightglow) originates in the earth's atmosphere. (From *Atlas of Representative Cometary Spectra* by P. Swings and L. Haser, Ceuterick Press, Louvain, 1956)

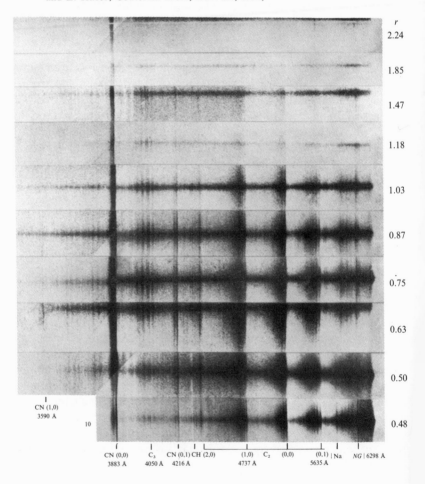

source for the coma and tail material could have been a single large body or a swarm of smaller bodies.

Occasional transits of comets across the sun's disk give an opportunity to search for a large nucleus. The Great Comet of 1882 transited the sun's disk and could not be seen. However, it was estimated that if the nucleus were larger than 70 km, it would have been seen. Similarly, Halley's comet was invisible when it transited the sun on May 18, 1910. Clearly, the nuclei of these comets could not be as large as 100 km.

Close approaches to the sun give an opportunity to estimate minimum sizes of at least some of the nuclear bodies. The Great Comet of 1882 passed sufficiently close to the sun to vaporize bodies about 1 meter in size. Although the comet showed some disruptions, its brightness did not change appreciably after perihelion. Similarly, the orbit of Comet 1843 I took it through the solar corona and within 0.1 solar radius of the photosphere without appreciable effect. The magnitude of the temperatures reached in comets at such close encounters is demonstrated by the fact that vaporization of dust takes place, as evidenced by the appearance of emission lines of metals in the comets' spectra. Thus, the nuclei of these comets could not have been as small as 1 meter.

The conclusion of the historical evidence was that a cometary nucleus consisted of a concentrated swarm of meteoroids with sizes in the range 10^{-4} cm to 10 meters. The strongest evidence for this conclusion was the identification of the orbits of several meteor streams with the orbits of specific comets. This concept is called the *sand* or *gravel bank model* of a comet.

The mass of this nuclear aggregate was – and still is – very difficult to measure. Dynamic determinations, if available, would be the most satisfactory. Unfortunately, comets have such low masses that they have never been observed to produce a measurable perturbation on another solar system body. For instance, on July 1, 1770, Lexell's comet passed within 2.3 million km of earth without producing as much as a 1-sec change in the length of the year. In 1805, Laplace demonstrated that the mass of the comet could not exceed 1/5000 that of the earth $(M_\oplus)$ or 4.6×10^{23} g. Calculations based on mutual perturbations of the two pieces of Biela's comet after it split in 1846 give a rough mass of $4.2 \times 10^{-7}M_\oplus$ or 2.5×10^{21} g.

Calculations on the Perseid meteor stream led B. Vorontsov-Velyaminov in 1946 to estimate a mass for Comet Swift–Tuttle of 5×10^{16} g. He also estimated a mass for Halley's comet in 1910 of 3×10^{19} g from the brightness of the reflected Fraunhofer spectrum; similar estimates were made by several others. If we assume a density of 2 as a reasonable value for most comets (50% silicates + 50% water ice), then the mass estimates, which depend on the cube of the radius of the nucleus, are sensitive to nuclear size estimates. Recent estimates place the nucleus of Halley's comet at about 2.5 km, yielding a mass of about 10^{17} g. Given the magnitude of the uncertainties, we conclude that the mass of Halley's comet is $10^{18\pm1}$ g and that of an average-size comet is $10^{17\pm1}$ g.

Now how does this relatively low-mass nuclear aggregate of rocks or swarm of meteoroids produce all the cometary phenomena usually observed? The historical answer was based on the variation of cometary brightness. The geometrical circumstances of comet observations are given in Figure 2.6. The law of cometary brightness can be written generally as

$$J = J_o f(\Delta) F(r) p(\cos \Theta) \tag{2.1}$$

where the brightness is in units of ergs per centimeter squared-second-frequency interval, Δ is the earth–comet distance, and $p(\cos \Theta)$ is the phase function. Significant phase effects in brightness are not widely accepted, even though comet observations have been made in a variety of geometrical circumstances, and, hence, $p(\cos \Theta)$ was set equal to 1. The geocentric variation $f(\Delta)$ was set equal to Δ^{-2}, because all evidence indicates that the only effect of the earth–comet distance is the usual inverse square law of brightness versus distance.[5] Thus, one obtained

$$J = \frac{J_o}{\Delta^2} F(r) \tag{2.2}$$

The function $F(r)$ represents the heliocentric distance variation of cometary brightness, a quantity that could well be governed by many factors. Hence, the discussion proceeded empirically by assuming $F(r) \propto r^{-n}$ to yield

Figure 2.6. Geometrical quantities used in equation 2.1.

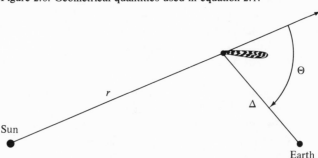

$$J = \frac{J_o}{\Delta^2 r^n} \qquad (2.3)$$

Cometary brightnesses were usually expressed in magnitudes, in terms of which equation 2.3 became

$$H = H_o + 5 \log \Delta + 2.5n \log r \qquad (2.4)$$

This equation was simplified further by defining $(H - 5 \log \Delta) = H_\Delta$, the magnitude reduced to earth. At length, one found

$$H_\Delta = H_o + 2.5n \log r \qquad (2.5)$$

According to equation 2.5, a plot of total cometary magnitude versus $\log r$ should have produced a straight line with slope $2.5n$. This simple formula provided (and still provides) a reasonable representation of the data, with n being the variable from comet to comet. Common individual values have been found to range between 2 and 6, with rare cases in the range -1 to 11. If the amount of reflecting or scattering material in the comet (rocks, dust, and gas) remained constant as it changed heliocentric distance, n would be 2. However, the average value was more like 4, which indicated a generation of material as the comet neared the sun. The material generated was assumed to be principally gas. In the 1940s B. U. Levin attempted to develop a physical picture of the relevant processes. To understand Levin's work, we must look at the physics of adsorption.

A solid material is known to cover itself with a monomolecular layer of gas or liquid that can be strongly attached in a process called *adsorption*. The outside surface that is covered includes all irregularities and cracks. The layer attached to the solid body is called the *adsorbate*, and the solid itself is called the *adsorbent*. The process whereby molecules are released from the surface of the solid body is called *desorption*. The processes of adsorption and desorption form the basis of Levin's physical picture. The processes are complex and depend on the nature of both the adsorbate and the adsorbent. The rate of desorption from a completely covered surface can be written as

$$\nu = s/\tau \qquad (2.6)$$

where s is the monolayer capacity of the adsorbent, which is defined as the surface concentration at complete covering, and τ is the average lifetime for a molecule to remain in the adsorbate. The lifetime is given by the empirically tested relationship known as *Frenkel's equation*, namely

$$\tau = \tau_o e^{+L/RT} \tag{2.7}$$

where L is the heat of adsorption (in calories per mole), T is the temperature (K), τ_o is the vibration period of a molecule attached to the surface, and R is the gas constant [2.0 cal $(K\text{-}mol)^{-1}$]. Because the principal temperature variation in equation 2.7 is in the exponential term, the desorption rate is roughly

$$\nu \propto e^{-L/RT} \tag{2.8}$$

If one assumes the brightness of the comet's head to be proportional to the desorption rate, then

$$J \propto e^{-L/RT} \tag{2.9}$$

To complete the derivation, we need to know the temperature of the adsorbent at different distances from the sun. This was supplied in 1929 by Herman Zanstra, who first derived the temperature of solid cometary particles assuming an equilibrium between absorbed and reemitted solar radiation. Zanstra's equation gives

$$T = \frac{T_o}{r^{1/2}} = \frac{289}{r^{1/2}} (K) \tag{2.10}$$

This equation[6] might not hold for particles larger than 1 cm if they keep the same side toward the sun. Thus the brightness of the head becomes

$$J \propto \exp(-Lr^{1/2}/T_oR) \tag{2.11}$$

or in magnitudes,

$$H_\Delta = 2.5 \log J + \text{const} = \frac{Lr^{1/2}}{RT_o} + \text{const} \tag{2.12}$$

At $r = 1$, $H_\Delta = H_0$, and we rewrite the equation as

$$H_\Delta = H_0 + \frac{L}{RT_o} (r^{1/2} - 1) \tag{2.13}$$

Equation 2.13 can be compared with the empirical relation (equation 2.5),

$$H_\Delta = H_0 + 2.5n \log r \tag{2.14}$$

At first sight, equations 2.13 and 2.14 might appear to be incompatible, but they are not. Over the range of r usually associated with cometary observations, $\log r \approx 0.87 (r^{1/2} - 1)$.[7] Hence, Levin's picture was found to be compatible with the observations. In addition, the empirically determined value of n is approximately related to the heat of desorption L by

$$n \approx 0.46 \; \frac{L}{RT_o} \qquad (2.15)$$

The observations determined a large range of values for L with an average value of 6000 cal mol^{-1}. Typical values measured in the laboratory were near 10^4 cal mol^{-1} and, hence, the cometary adsorbate was rather volatile by terrestrial standards.

However, Levin's theory had at least one fatal flaw. Laboratory studies showed that 1 g of meteoritic material was capable of supplying a total of 10^{19} molecules, a number that included both adsorbed molecules (discussed above) and molecules that had penetrated well into the solid and, accordingly, required more energy for extraction than the surface-layer molecules. For a comet with a typical nuclear mass of 10^{18} g, the total gas supply available could not exceed about 10^{37} molecules. However, Wurm has shown that the typical comet contains approximately 10^{35} or 10^{36} molecules in its coma, and these molecules leave through the cometary tail in 1 day or less. Thus, the total meteoritic gas supply would maintain the supply of molecules observed in comas of comets for a period of time ranging from 10 days to at most a few months. Of course, we know comets can have comas visible for many months, and some of the periodic comets have been seen on numerous returns. Thus, although the adsorption–desorption picture is compatible with the relative rate at which gas is released, it falls far short of explaining the total amount of gas released. As we shall see below (Chapters 4, 5), the solution to the problem is not to produce the gases from a minor impurity, but from the material in the nucleus itself. Ices provide a suitable material, and they can generate an ample supply of gases to maintain comets: 10^{22}–10^{23} molecules g^{-1}.

The origin of comets

The nineteenth century saw considerable speculation on the origin of comets. Orbital information rather than the physical state of the comet was the basis of these efforts.

As Chambers would write in 1910:

Two provisional answers suggest themselves: either (1) comets are chance visitors wandering through space and now and again caught up by the Sun, or by some of the major planets . . . and compelled to attach themselves to the Sun and by taking elliptic orbits to become permanent members of the solar system; or (2) they are aggregations of primaeval matter not formed by the Creator into substantial planets, but left lying around in space to be picked up and gathered into entities as circumstances permit. [Olivier, 1930:208]

Also in 1910, A. C. D. Crommelin wrote concerning the origin of comets:

(1) that they are the products of eruptions from the sun; (2) that they are the product of eruptions from the larger planets in a sunlike state; (3) that they are stray fragments of the nebula which is supposed to have been the parent of our system, and that they remained unattached to any of the large masses that were formed from the nebula. [Proctor and Crommelin, 1937:179]

Theories based on an interstellar origin and subsequent capture within the solar system have been advocated by Laplace (in 1813), Heis, Schiaparelli, von Seeliger, Fabry, Bobrovnikoff, and Nölke. A theory of purely interstellar origin encountered insuperable difficulties basically because we did not observe comets with hyperbolic orbits. In 1860 the renowned solar physicist R. C. Carrington noted that comets originating outside the solar system would as a rule have hyperbolic orbits. He also noted that as the solar system moved through the supposed interstellar cloud of comets, we should observe more comets coming from the direction of the solar motion than from the opposite direction. Any doubt concerning the lack of hyperbolic orbits was removed by Ellis Strömgren in 1914 when, by considering planetary perturbations, he showed that (within the errors of observation) there was no evidence from observations that any comet entered the solar system with a hyperbolic orbit. In addition, Henry Norris Russell in 1920 showed that planets were not efficient in "capturing" comets. Only Jupiter has a legitimate comet family. Saturn might have a comet family, but Uranus and Neptune certainly do not. The capture process, even if the comets with hyperbolic orbits came long ago, is basically inefficient and cannot account for the observed orbits of comets. Subsequent research by E. Strömgren confirmed Russell's conclusions.

The capture theory might be revived in two ways. One way is the theory proposed by R. A. Lyttleton (1953) in which the condensations in the interstellar medium are caused by the passage of the sun. These condensations travel in hyperbolic orbits with respect to the sun and intersect each other on the line representing the solar motion in the antiapex direction. Collisions of these condensations were thought to produce, according to Lyttleton, the sand or gravel bank model of comets. Because the collisions would be inelastic, the hyperbolic motion could be changed into the elliptical and parabolic motion observed. There are many difficulties with Lyttleton's theory, and it has very few advocates at present.

The second way to save the capture hypothesis would be to have comets originate in an interstellar cloud that traveled at the same speed and direction as the solar motion. Such a hypothesis is highly artificial and has probably never been seriously proposed. A perfectly natural way to achieve the desired result is to have both the solar system (sun and planets) and the comets formed out of the same interstellar cloud.

This is the same basic idea proposed above by Chambers (second suggestion) and by Crommelin (third suggestion).

However, the fact that comets had to be solar system objects gave support to other theories of their origin. Crommelin seemed inclined toward a solar origin at least for some comets, such as the sun-grazing group of comets that included Comet 1882 II. Material was certainly observed to be ejected from the sun at great velocities, and this fact supported the idea. However, it seemed impossible to explain the origin of comets with perihelion distances well away from the sun. The solar hypothesis also suffers from another problem: the fact that condensation of ejected solar material into comets seems highly unlikely. Fortunately, most other theories of cometary origin are not encumbered with this objection.

The final historical theory of cometary origin within the solar system ascribes comets to ejections from the major planets, principally Jupiter. Through the years Lagrange,[8] Proctor, Tisserand, and Vsekhsvyatskij have advocated a planetary origin via ejection. S. K. Vsekhsvyatskij envisoned a volcanolike process in his earlier work (cf. Vsekhsvyatskij, 1977), but the large escape speed from Jupiter (67 km sec^{-1}) seemed too high to be attained by known physical processes. In his later work the site of origin was transferred to the satellites and rings of the major planets, where the escape speed is lower by a factor of 10. The near-parabolic orbits were supposed to have been generated by ejection from the outer planets, starting with Uranus and including hypothetical trans-Plutonian planets.

An alternate proposal places the site of ejection of comets on the satellites of the major planets. The satellites thought by Vsekhsvyatskij to be responsible are Europa and Callisto (Jupiter), Titan (Saturn), Titania (Uranus), and Triton (Neptune). The Galilean satellites of Jupiter were examined closely in 1979 by the *Voyager* spacecraft. Intense volcanic activity has been found on Io; however, its density leads to a model of mostly silicate material. Europa and Callisto probably have icy crusts; however, no sign of volcanic activity was found. At least for these three Jovian satellites, the volcanic-type ejection of compact cometary bodies seems very unlikely. The most telling argument against the major planets as a source of comets is the observed distribution of $1/a$ values, with a peak between 10^{-4} and 10^{-5}. Perturbations acting on comets arising in these outer planets cannot produce such a distribution.

Comet observers and comet observations

No discussion of the history of comets would be complete without some mention of the famous comet hunters and observers.

Among the most celebrated discoverers were Charles Messier (1730–1817), who discovered 21 comets, Caroline Herschel (1750–1848) with 8, Lewis Swift (1820–1913) with 11, Giovanni Donati (1826–73) with 6, William Brooks (1844–1921) with 20, and E. E. Barnard (1857–1923) with 19. There is considerable confusion concerning the number of comets discovered by some of the individuals listed; hence, the numbers given are only approximate.

The grand champion is Jean Pons (1761–1831), who is generally credited with the independent discovery of 37 comets; because communications at the time were slow and often difficult, some of the comets were independently discovered by others. Pons began his career as the concierge of the Marseilles Observatory, and his reputation earned him the nickname of the Comet's Magnet. He searched for comets with telescopes figured and constructed entirely by himself; one of these had a field of about 3° and was called the Grand Chercheur. Unfortunately, Pons's descriptions of comets' positions were not accurate. His name lives through Comet 1819 III, which is now called *Comet Pons–Winnecke*. But Pons's most famous discovery does not bear his name; he discovered Encke's comet in 1805 and again in 1818. Encke himself always referred to it as the *Comet of Pons;* fortunately, Pons had so many comet discoveries to his credit that one could be spared.

Charles Messier was a keen comet searcher, but we remember his name for his catalogue of nebulous objects. He compiled the catalogue to avoid being misled by celestial objects that could be mistaken for comets, and so he introduced his famous M numbers into the astronomical language. His enthusiasm for comet discoveries earned him the nickname of the Ferret of Comets by King Louis XV of France.

The spirit of these early comet observers and discoverers was carried well into the twentieth century by Georges Van Biesbroeck (1880–1974). He discovered three comets and is listed as the recoverer of many periodic comets, including Encke and Pons–Winnecke.

In the nineteenth century, the discovery of comets was encouraged by the offering of medals and cash prizes. For instance, the King of Denmark offered a gold medal and an award for such a discovery. Among the historically interesting recipients of this medal was Maria Mitchell of Nantucket Island for her discovery of Comet 1847 VI. Later in the century, a prize of $200 was offered by H. H. Warner for comet discoveries. E. E. Barnard won this prize several times and used the money to pay for his home in Nashville, Tennessee. For many years, beginning in 1890, the Astronomical Society of the Pacific awarded the Donahue Medal for the discovery of a comet.[9]

Barnard's observations of comets were important, particularly his work on Comet Morehouse in 1908. He is also credited with the first discovery of a comet by means of a photographic plate on October 12, 1892. His photograph of Comet 1892 V showed traces of a tail. This credit excludes the "eclipse comet" seen only on photographs of the solar corona taken at the total solar eclipse on May 17, 1882.

Photography of comets began with Donati's comet in 1858. An English commercial photographer (Mr. Usherwood) obtained a small overall view of the comet showing the tail, using a 7-sec exposure made with an $f/2.4$ camera. Two days after Usherwood's success, astronomer G. P. Bond (1825–65) obtained a photograph that showed only the nucleus with a 360-sec exposure through an $f/15$ lens.

Completely successful photographs were obtained for Tebbutt's comet (1881 III). The success was partly due to the comet's exceptional brightness and the development of dry (silver bromide) plates. On June 30, 1881, P. J. C. Janssen (1824–1907) obtained a beautiful photograph (half-hour exposure) showing approximately 2.5° of tail (see Figure 2.7). Henry Draper also obtained a photograph of this comet (exposure time: 2 hr 42 min) showing approximately 10° of tail. This comet was also the first to have its spectrum recorded photographically.

The bright comet of 1881 was followed by another in 1882. Among its observers was David Gill (1843–1914) at the Cape of Good Hope Observatory. He used an $f/4.5$ camera borrowed from a local photographer and mounted it on the observatory's equatorial telescope. The photographs obtained had exposures in the range 20 min to 2 hr. These

Figure 2.7. Reproduction of one of Janssen's photographs of Comet 1881 III. (From Annuaire du Bureau des Longitudes, 1882; Courtesy of the Observatoire de Paris, Meudon)

photographs (see Figure 2.8) were important for two reasons. First, they confirmed that comets could be photographed easily with cameras or lenses of large angular aperture (small focal ratio); the linear aperture did not matter. From this time on, comets would have their pictures taken at several observatories around the world whenever they appeared. Second, the photographs showed a large number of faint stars. This fact convinced Gill that large-scale stellar photography was entirely practical. He decided to use photography to obtain the Map of the Southern Heavens. Under Gill's direction the plates were taken by R. Wood. The measurements (for precise position and approximate magnitude) were carried out under the direction of J. C. Kapteyn. This effort produced the first great modern catalogue, the *Cape Photographic Durchmusterung* (1896–1900).

The modern era: an introduction

We will say that the modern era of cometary astronomy began in 1950. In that year the first of Whipple's papers on the icy-conglomerate model of the cometary nucleus and Oort's paper proving

Figure 2.8. The great comet of 1882 as photographed by Gill in November 1882. (Courtesy of the Royal Astronomical Society)

the existence of a vast reservoir of new comets were published; hence it is a natural (but, of course, not rigid) boundary in time. In addition, as we will presently establish, many other lines of investigation that are important to the understanding of cometary phenomena began at approximately the same time. The description of some items in this introductory section will be brief, particularly when they are discussed in more detail in subsequent sections.

The icy-conglomerate model of the cometary nucleus was originally presented by Whipple in three papers beginning, as noted above, in 1950. In this model the nucleus is presented as a rather large mass of ices such as H_2O, NH_3, CH_4, CO_2, and C_2N_2, with particles (sizes from ~ 1 μm to meteoroids) embedded throughout. In addition, the structure has enough tensile strength to resist tidal disruption by a close encounter with a planet or the sun. The ices postulated by Whipple also provide an adequate supply of gas for the comet because each gram of ices yields 10^{22}–10^{23} molecules. Thus, for equal masses, ices are a better source of gas than are meteoritic grains or stones by a factor of 10^3 or 10^4. However, it is essential that the ices exist in a single structure or perhaps a small number of structures; very small isolated ice grains would have a short lifetime (see Chapter 4). Ices are generally poor conductors of heat and, thus, the incident solar radiation heats only a surface layer that is slowly eroded away by sublimation.

The dust (~ 1 μm size) observed in cometary comas and type II tails is released when the ices sublimate. Some of the dust could remain attached to the nucleus to form an insulating structure.

If the nucleus rotates, the *rocket effect* (see the suggestion by Bessel, Chapter 1) can operate to produce the nongravitational forces required to explain the orbits of some comets, such as Encke. Heat is supplied to the nucleus by solar radiation, and it takes time for the heat to penetrate to the icy layers and to cause sublimation. Thus, the evaporation of the gas molecules (which form the coma) takes place in a preferential direction that makes an angle with the sun–comet line (see Figure 2.9). The recoil due to this preferential, nonisotropic mass loss produces the nongravitational forces on comets.

Thus, Whipple's icy-conglomerate model can supply the gases known to come from comets at many apparitions, explains the nongravitational forces, supplies the dust observed in type II comet tails, and has tensile strength to resist tidal disruption. Eventually, all of the icy material is sublimated, and the remaining dust and stones are spread out by perturbations along the comet's orbit to form meteor streams. The model has roughly three-fourths of the mass in ices and the remainder in meteoritic dust and stones. Some of the dust particles emitted by active comets would be spread throughout the solar system by planetary perturbations and the Poynting–Robertson effect. Whipple concluded that the cometary source was sufficient to supply the zodiacal light indefinitely.

The icy-conglomerate model even held hope of explaining the jetlike phenomena so frequently observed near cometary nuclei. The nucleus could be rather inhomogeneous, and packets of gas might be able to develop. Donn and Urey in 1956 extended the icy-conglomerate model to include explosive chemical reactions involving possible free radicals or unstable molecules. The explosions would contribute to the activity observed in comets, including jets and outbursts.

Donn and Urey also commented on the chemistry of Whipple's model. The presence of methane (CH_4) and carbon dioxide (CO_2) in the same object seemed implausible because in methane, carbon is in its most reduced state, whereas in carbon dioxide it is in its most oxidized state. Because carbon dioxide (in the form of CO_2^+) was observed in comet tails, there was the possibility that methane was not an important constituent of the cometary nucleus.

The most significant problem with Whipple's original icy-conglomerate model was the rate of production of gas at different heliocentric distances, particularly the usual onset of gaseous activity at approximately 3 AU. Very distant comets usually[10] show only a continuous spectrum. Near 3 AU the CN spectrum appears, and near 2

Figure 2.9. Schematic of the rocket effect.

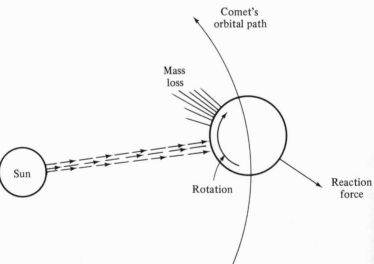

AU the spectra of C_3 and NH_2 appear. As the comet continues inward and approaches 1.5 AU, the spectra of C_2 (Swan bands), CH, OH, and NH appear. These spectra grow stronger as the comet moves inside 1.5 AU and the spectra of CO^+, OH, N_2^+, and CH^+ appear in the tail. The lines of Na appear near 0.8 AU and, if the comet's perihelion distance is very small (≈ 0.1 AU), lines of Fe, Cr, and Ni appear. This sequence is an averaged idealization that may not apply to a particular comet. Originally, it was thought that the spectroscopic sequence could be explained by the more volatile molecules evaporating at the larger heliocentric distances, coupled with the subsequent dissociation of various relatively complex parent molecules. However, when examined in detail, this picture could not satisfactorily account for the onset of gaseous activity near 3 AU.

The onset of activity appeared compatible with the sublimation of water ice; that is, a hypothetical frozen H_2O comet approaching the sun would begin gaseous activity near 3 AU. Water is not easy to detect from the ground and, hence, the spectra observed (CN) could result from other constituents released at the same time. Delsemme and Swings in 1952 suggested that the nucleus was composed of solid hydrates of the species thought to be needed, such as $CH_4 + 6H_2O$, and so on. The picture based on solid or clathrate hydrates has been presented by Delsemme and his colleagues in a series of papers (Delsemme and Wenger, 1970; Delsemme and Miller, 1970, 1971a,b). Some substances form crystal structures that have cavities large enough to permit occupancy by noble gas atoms or molecules. Water molecules can form a hydrogen-bonded framework with cavities that are polyhedra with pentagonal or hexagonal faces. An example for xenon is shown in Figure 2.10. Each unit cube contains 8 xenon atoms and 46 water molecules; thus, the clathrate is essentially a hydrate, $Xe-5\frac{3}{4} H_2O$. Similarly, methane forms a clathrate hydrate, $CH_4-5\frac{3}{4} H_2O$.

Delsemme's introduction of clathrate hydrates into Whipple's icy-conglomerate model appears to have solved the problem of the onset of gaseous activity in comets near 3 AU. The essential point is not the existence of the exact crystal structure of the clathrates but the existence of cavities in which other molecules can be trapped. The rest of the spectroscopic sequence results from excitation phenomena. The sodium and metals lines arise from the vaporization of dust as the comet nears the sun.

Delsemme's model is compatible with other cometary phenomena. For example, the sublimation rates are sufficient to produce the nongravitational forces required to explain orbital data in detail. Delsemme's model also produces an icy-grain halo (caused by the stripping of ice grains from a solid as the ice sublimates) that may reflect enough sunlight to produce the so-called false or photometric nuclei of comets. This model is discussed in detail in Chapter 5.

The view of water ices as the principal constituent of the nucleus received an additional boost in 1970 with the observation of huge clouds of atomic hydrogen around comets. Comet Tago–Sato–Kosaka (1969g) was observed in Lyman α (λ 1216) from the Second Orbiting Astronomical Observatory (OAO-2) by Code, Houck, and Lillie (1972). In April 1970, Lyman α radiation was measured from Comet Bennett by Bertaux and Blamont using photometers on the Fifth Orbiting Geophysical Observatory (OGO-5). Comet Bennett was also detected in Lyman α by the OAO-2 observers (see Figure 4.26) and by Thomas using the other Lyman α photometer on OGO-5. The impressive aspects of these discoveries were these: (1) they had been predicted in some detail by Biermann in 1968 and (2) the clouds of hydrogen had

Figure 2.10. Structure of a clathrate crystal, xenon hydrate. (From *General Chemistry,* 3rd ed., by L. Pauling, New York, W. H. Freeman, 1970)

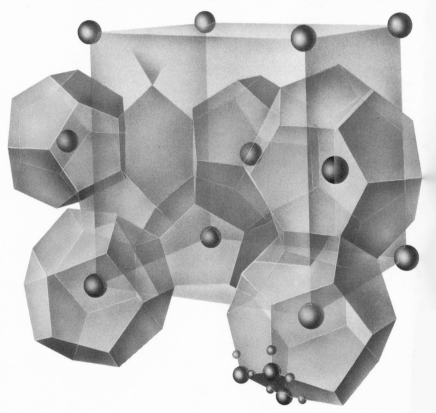

huge dimensions, some 1.5×10^7 km or about 0.1 AU. The clouds around these two comets were rather similar. In December 1970 the hydrogen cloud around Comet Encke was detected; this cloud had a dimension of around 10^6 km.

The spectrometer on OAO-2 obtained spectral scans of Comet Bennett over the range 1000 to 4000 Å (see Figure 4.26). The strong emission near 3090 Å is due to the OH radical. Its strength leads us to the conclusion that OH has an abundance comparable to that of H. The fact that both H and OH are present is strong evidence that they originated from the dissociation of water molecules. Furthermore, the large abundances of H and OH lead us to the conclusion that water is the principal constituent of the cometary nucleus. In 1962 Swings and Greenstein discovered the forbidden red oxygen line near 6300 Å in the spectrum of comets. The presence of this line, with its very low transition probability, indicates that large quantities of O exist in cometary atmospheres. One possible source of the O could be the dissociation of the OH radical; however, the dissociation of H_2O directly into $H_2 + O$ and the dissociation of CO_2 into $CO + O$ probably also contribute to the presence of O. The existence of O does point to a large abundance of OH in the cometary atmosphere, and the evidence is entirely consistent with the idea that H_2O is the major parent molecule for H, OH, and O.

This picture was strengthened and refined by observations of Comet Kohoutek. The H_2O^+ ion was identified by Herzberg and Lew (1974) by comparing laboratory spectra with cometary spectra taken by Herbig and by Benvenuti and Wurm. Evidence has also accumulated for large amounts of CO or CO_2 in cometary atmospheres. If the nucleus contained roughly 10–15% CO or CO_2 by number trapped in the clathrate-like ice cavities, the presence of CO in the halo and the predominance of CO^+ as the observed emission in type I tails would be explained. It seems likely, then, that the predominance of CO^+ emission in plasma tails is due to the large amount of CO present in comets rather than to ionization/excitation phenomena or to limited spectral coverage.

An important area of cometary physics was discovered by Biermann in 1951. He studied changes in the position of features in type I tails; knots or clouds of gas showed repulsive accelerations that are represented by the parameter $(1 - \mu)$ of 10^2 or more times solar gravity. As has been noted, such accelerations could not be produced by radiation pressure. Biermann attributed the accelerations to momentum transferred to the tail ions (largely CO^+) by a continuous outflow of ionized material from the solor corona, called the *solar corpuscular radiation*. Biermann developed a simple quantitative model for the momentum transfer – between a proton–electron stream and an ion–

electron gas. The acceleration on the ions was

$$\frac{dV_i}{dt} \approx \frac{e^2 N_e V_e m_e}{\sigma m_i} \approx 10^{-4.3} \frac{m_e}{m_i} N_e V_e \tag{2.16}$$

Here N_e and V_e are the electron density and the bulk velocity in the beam, respectively, m_e and m_i are the masses of the electrons and ions, respectively, and σ is the electrical conductivity that has been evaluated approximately.

For $V_e = 1000$ km sec^{-1} (from geomagnetic delay times) and an assumed $N_e = 600$ cm^{-3}, the acceleration is 100 cm sec^{-2}. At the orbit of earth, the solar gravity is 0.59 cm sec^{-2} and, hence, 100 cm sec^{-2} corresponds to a $1 - \mu$ of roughly 170, a value adequate to explain the observations. The problem with this solution was the assumed electron density of 600 cm^{-3}. Although this value was in agreement with other evidence at the time, it is now known to be about two orders of magnitude too high.

Biermann's picture was also supported by Hoffmeister's (1943) work on the orientations of plasma tails. As viewed in the plane of the comet's orbit, the plasma tails were not strictly radial but lagged behind the radius (in the sense opposite the comet's motion) by an aberration angle given by

$$\tan \epsilon \approx \frac{V_\perp}{w} \tag{2.17}$$

Here $V_\perp$ is the component of the comet's orbital velocity perpendicular to the radius vector and w is the velocity of the corpuscular radiation. The aberration angle ϵ is about 5° and corresponds to a w of about 450 km sec^{-1}. The lag occurs because a comet's motion through a medium produces a "wind" as seen by a hypothetical observer riding on the comet.

This aberration effect is probably the most durable early cometary evidence for the existence of the solar corpuscular radiation, or the *solar wind* as it is now called. These studies emphasized the facts that the solar wind shaped parts of comets and that a proper understanding of the solar wind interaction was necessary for a sound physical picture of comets. The study of plasma tail orientations remains a valuable method for inferring solar wind velocities (Chapter 6).

The interplanetary or solar wind magnetic field was introduced into the study of comet tails by Alfvén in 1957. If a plasma beam (protons plus electrons) with a frozen-in magnetic field encounters a comet neutral molecules that are ionized (perhaps by the encounter) find themselves attached to the field lines. This effect "loads" the field line near the head of the comet, but not those far away. Thus, the field line are hung up on the head of the comet, as illustrated in Figure 2.11, which shows the view perpendicular to the field of the beam; the view parallel to the beam is also shown in Figure 2.11. The two views are different unless there is sufficient turbulence.

The introduction of the magnetic field by Alfvén naturally explained the narrow, straight streamers that make up the type I tail. The magnetic field might enhance the coupling between the interplanetary and cometary plasmas and perhaps produce the accelerations of features observed in type I tails without the need for the high densities required in Biermann's essentially viscous picture. Hoyle and Harwit in 1962 investigated the possibility that the interaction in the field-free case

Figure 2.11. Interaction of the solar wind magnetic field, B, with a cometary ionosphere. Parts (a) through (d) are a time sequence as viewed perpendicular to the magnetic field. Part (e) is the view parallel to the field. (After H. Alfvén, 1957, On the Theory of Comet Tails, *Tellus 9*:92–6)

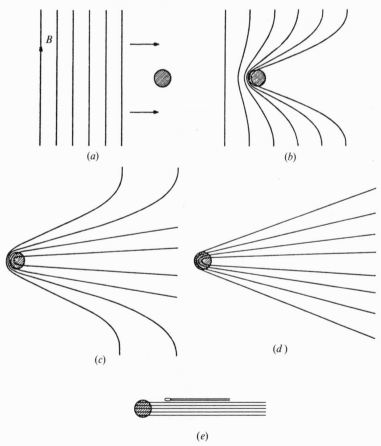

might be enhanced by collective effects; they concluded that plasma instabilities were not effective. A companion study by Harwit and Hoyle also in 1962 showed that capture of an interplanetary magnetic field by a comet could lead to a coherent picture. The compression of field lines in the head could squeeze plasma out of a magnetic tube of force (much like toothpaste out of a tube) and thus account for the initial velocity of tail material of roughly 20 km sec^{-1} near the head. After the material enters the tail proper, it essentially expands into a medium with zero pressure at large distances from the head. Plausible pressure differences could produce the accelerations observed for the tail knots and clouds.

The question of cometary orbits and the origin of comets was also under active study during our modern era. In fact, some of the speculation about the origin of comets has at times outstripped the necessary parallel progress in the physics of comets. This chapter of our story also begins in 1950 with a study by Oort. The idea of a very large cloud of comets surrounding the inner solar system has been known for some time. Öpik had discussed the idea in 1932, and Young's *General Astronomy* ascribed the idea of a comet-dropping envelope a very large distance from the sun to Peirce in the nineteenth century. But in 1950 Oort developed the idea that has become, with modifications, the most widely accepted view of the origin of comets. Oort's investigation was stimulated by van Woerkom's (1948) study on the origin of comets. The paper by van Woerkom contains many items of interest, in particular, the idea that comets diffuse inward from a field of parabolic comets with Jupiter's perturbations causing the diffusion.

The basic data of Oort's paper were the distribution of original values of $1/a$ (a being the semimajor axis in astronomical units); the "original" values refer to the time when the comets were still well outside the orbits of the major planets. The 19 values for which the mean error in $1/a$ is less than 0.00010 came from the work of E. Strömgren. Oort interpreted the frequency curve implied by Table 2.1 as indicating a steep maximum for very small values of $1/a$ and concluded that a substantial fraction of these comets originated in a region of space extending from 20,000 AU to at least 150,000 AU from the sun. Although this distance extended almost to the nearest star, Oort noted that the comets were not interstellar because of the absence of hyperbolic original orbits; hence, he inferred that they were always members of the solar system. But how could these comets be sent into the inner solar system where we observe them? Oort considered the only viable mechanism to be perturbations caused by passing stars, and he studied that mechanism in detail. He concluded that the cloud and stellar per turbations could account generally for the observed distribution curve of $1/a$ as well as for the random distribution of orbital planes and perihelia, and for the preponderance of nearly parabolic orbits. Oort estimated that the cloud contained roughly 10^{11} comets with a total mass between 1/10 and 1/100 the mass of the earth.

The short-period comets could be accounted for by the diffusing action of Jupiter; van Woerkom's calculations indicated an average change in $1/a$ caused by a comet's passage through the inner solar system of 0.0005, a change that could be either positive or negative. This provides the input to the short-period comets. They are lost by disintegration in the inner solar system and by ejection due to planetary perturbations.

Oort also speculated that comets were formed at an early stage of the planetary system from the asteroid ring and "brought into large, stable orbits through the perturbing actions of Jupiter and the stars" (Oort, 1950). Meteoroids were thought to have a common origin.

Finally, Oort offered an explanation for the fact that about 20 times more comets with very small values of $1/a$ were observed than he would predict. This excess of "new comets" was "taken to indicate that comets coming for the first time near the sun develop more extensive luminous envelopes than older comets" and were thus more likely to be discovered. He noted that this phenomenon might warrant additional study. Oort and Schmidt in 1951 studied the differences between old and new comets. Comets entering the inner solar system for the first time show a stronger continuous spectrum, presumably because they are "dustier." The dust could be liberated as the volatiles near the surface evaporate. On subsequent passes through the inner solar system, the volatiles near the surface are not available, and there is less

Table 2.1. *Distribution of original values of* $1/a$

$1/a$ (AU)	n
≤ 0.00005	10
0.00005–0.00010	4
0.00010–0.00015	1
0.00015–0.00020	1
0.00020–0.00025	1
0.00025–0.00050	1
0.00050–0.00075	1
> 0.00075	0

evaporation and dust release. The large dust release for new comets makes them anomalously bright and accounts for their enhanced rate of discovery.

Other orbit-related processes (besides the origin of comets) have been under investigation during this period. In a series of papers Marsden and associates have investigated the nongravitational forces on comets. Initially, they assumed reasonable forms for the necessary functions (because the relevant physics was uncertain), but their recent papers use functions based on Delsemme's vaporization temperature of an icy conglomerate controlled by the latent heat of water snow. Reasonable values of variables allow the construction of orbits based on both nongravitational and gravitational forces; these give a good representation when compared with the observed positions of comets. The radial nongravitational force is usually about 10 times the transverse component in the plane of the comet's orbit. Thus, the net force makes an angle of about 6° with radial direction. However, it is the transverse force that is cumulative and produces most of the observed effects. This force can act either in the direction of the comet's motion or in the opposite direction. The nongravitational force also changes with time. Results for Comet Encke show a maximum in the transverse nongravitational force near 1820, a fact that may have contributed to the discovery of nongravitational forces at that time. Marsden and associates have also considered the influence of nongravitational forces on the calculation of original orbits. The nongravitational forces introduce substantial errors into the original values of $1/a$ and, in addition, a systematic effect that reduces the radius of the outer part of the Oort cloud to approximately 50,000 AU.

A statistical approach to the capture of short-period comets by Jupiter has been undertaken by Everhart (1972). His basically numerical approach, which utilizes the Tisserand criterion (equation 3.10), allows him to follow the orbits of tens of thousands of "comets" for up to 2000 returns per "comet." Everhart finds that the cumulative effects of Jupiter's perturbations can produce comets with orbital properties closely resembling those of the short-period comets. However, only comets with small inclinations and perihelia near the orbit of Jupiter are involved. This means that the short-period comets result from a sample of comets not normally observed from earth. Although some of the details have been questioned, the general approach adopted by Everhart seems very promising.

Finally, late 1973 and early 1974 saw an unprecedented effort toward observations of a comet, namely, Comet Kohoutek. Spacecraft and astronaut observations of this comet, coupled with an enormous ground-based effort, have given us an invaluable fund of new information

tion. These new data are discussed in more detail later (Chapter 7). Here we mention that the evidence for water as the prime constituent of the nucleus has been strengthened and that evidence for carbon monoxide or carbon dioxide as the principal secondary constituent is beginning to emerge. In addition, Comet Kohoutek probably has been detected in the radio region in both line and continuum radiation. For example, methyl cyanide, CH_3CN, a molecule known in the interstellar gas, was detected, but with a low signal-to-noise ratio.

Comet Kohoutek has greatly enhanced our understanding of cometary physics. Other bright comets of the 1970s have contributed to our knowledge. It is almost trite to point out that the picture we have built up is subject to revision in the future. The event that is filled with potential for enormous advances in cometary science is the in situ investigation of comets (Chapter 9). The National Aeronautics and Space Administration (NASA) has a cometary mission on the drawing board. The target comet or comets for a mission have not yet finally been chosen. However, a highly likely scenario involves a spacecraft that would fly by Halley's comet on a near-collision course, dropping a probe into the vicinity of the nucleus as it passed, and then proceed to Comet Tempel 2 and fly in formation with that comet for many months. The mission would provide close-up pictures of cometary phenomena such as the nucleus and would permit direct sampling of the material in the coma. Such a mission may be the only means for us to ascertain the nature of a cometary nucleus. A strong motivating force for the in situ study of comets is the widely held belief that as remnants of the solar system's formation, they are the most pristine, primitive objects still to be found in the solar system.

Part II

The current perspective

In Part I we have seen how our modern conception of a comet has developed, and we have described that conception very briefly. The picture is simple at one level. A comet is a chunk of ice – mostly water, but other substances as well – heavily laced with meteoroids. If comets were isolated in space, they would be relatively uninteresting. However, comets are part of the solar system and as such are subjected to interactions with solar electromagnetic and corpuscular radiation. They are also subjected to the gravitational influence of the sun, planets, and, perhaps at some stage in their existence, of another star. It is the phenomena of these interactions that make comets interesting.

In Part II we will take a detailed look at cometary phenomena. We will stress the physical processes by which comets interact with their environment, and in doing so we will understand more fully how the modern picture has been inferred from the observations.

Halley's comet as it approached and receded from the earth in 1910. (Hale Observatories)

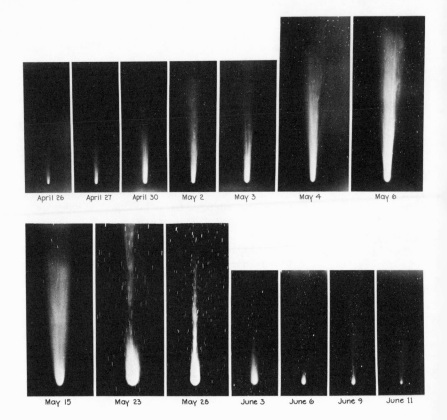

3   *Dynamics of comets*

The discovery of comets

Each year roughly 10 comets are discovered or rediscovered. To date about 1000 comets are known, 400 of which were discovered before the telescope was invented. *The Catalogue of Cometary Orbits* published in 1979 by Marsden gives orbital parameters for 1027 cometary apparitions, which result from approximately 658 individual comets. Some 545 of these are called *long-period comets,* a group of comets that have periods greater than 200 years. Roughly 113 comets are periodic or short-period; they have periods of less than 200 years. Multiple returns of the periodic comets account for the remainder in the catalogue.

Comets continue to be discovered by both amateur and professional astronomers. The amateur astronomers search the sky systematically using wide-field, low-magnification telescopes or binoculars. A number of the brightest recent comets have been discovered by amateurs. An example is Comet Ikeya–Seki, which was a naked-eye comet in 1965; it was discovered by two well-known, quite successful Japanese amateurs. Comets discovered by professionals are often the fainter ones, which are frequently recorded by accident on wide-field plates taken for other purposes. Luboš Kohoutek discovered two comets in 1973 while he was taking plates to study asteroids. Of these two comets, one remained very faint and the other reached naked-eye brightness. The discovery of a comet is communicated to the Central Bureau for Astronomical Telegrams, Smithsonian Astrophysical Observatory (Cambridge, Massachusetts) and then announced by an International Astronomical Union telegram, an example of which is shown in Figure 3.1.

Ordinarily, comets are named after their discoverer – or codiscoverers if several nearly simultaneous reports are received. However, a few comets have been named after the persons who computed their orbits. Well-known examples are Halley's comet, Encke's comet, and, more recently, Crommelin's comet.[1]

The names of some multidiscoverer comets can be cumbersome. Two shorter designation systems have also been devised. When a comet is discovered, it is designated by the year of discovery and the order of discovery in that year. Thus, the first comet discovered in 1978 was called 1978a, the second 1978b, and so on. After the orbits of the

59

comets are calculated and their time of perihelion passage is ascertained, they are designated according to the year and order of perihelion passage. The first comet to pass perihelion in 1978 was 1978 I, the second 1978 II, and so on. Unfortunately, confusion can exist in these designations, especially if improved orbit calculations change the predictions of order of perihelion passage or the year.

Figure 3.1. An International Astronomical Union announcement telegram.

Circular No. 2511

CENTRAL BUREAU FOR ASTRONOMICAL TELEGRAMS

INTERNATIONAL ASTRONOMICAL UNION

POSTAL ADDRESS CENTRAL BUREAU FOR ASTRONOMICAL TELEGRAMS
SMITHSONIAN ASTROPHYSICAL OBSERVATORY CAMBRIDGE MASS 02138 USA
CABLE ADDRESS SATELLITES NEWYORK · WESTERN UNION RAPID SATELLITE CAMBMASS

COMET KOHOUTEK (1973f)

Dr. L. Kohoutek, Hamburg Observatory, Bergedorf, telexes that he has discovered another new comet, as follows:

1973 UT	α_{1950}	δ_{1950}	m_1
Mar. 7.86944	$8^h22^m.4$	$+ 4°31'$	16
9.85972	8 21.0	$+ 4 40$	16

The comet is diffuse, with central condensation, no tail.

COMET KOHOUTEK (1973e)

The following precise positions have been reported:

1973 UT	α_{1950}	δ_{1950}	m_1	Observer
Mar. 2.73681	$10^h58^m00^s.30$	$+25°05'32''.5$	15	Kojima
2.74827	10 57 57.24	$+25 06 31.8$		"
7.72014	10 32 29.05	$+31 34 31.1$	15	"
7.73472	10 32 24.48	$+31 35 36.5$		"
7.84688	10 31 47.78	$+31 44 08.4$		Kohoutek
7.85590	10 31 44.73	$+31 44 48.3$		"
9.87431	10 20 21.62	$+34 14 55.1$		"
9.88403	10 20 18.32	$+34 15 37.2$		"
10.63854	10 15 54.27	$+35 10 01.2$	15	Kojima
10.69966	10 15 32.48	$+35 14 23.2$		"

N. Kojima (Ishiki). Measurer: H. Kosai. Communicated by K. Osawa. The Mar. 2.7 position on IAUC 2506 is incorrect.
L. Kohoutek (Hamburg Observatory). Nuclear magnitude estimates are: Feb. 28.0, 14.5; Mar. 7.8, 14.7; Mar. 9.9, 14.7.

COMET HECK-SAUSE (1973a)

Further precise positions have been reported as follows:

1973 UT	α_{1950}	δ_{1950}	m_1	Observer
Jan. 26.92152	$12^h07^m32^s.58$	$+21°54'58''.2$		Ferreri
26.92708	12 07 32.08	$+21 55 11.7$		"
30.01388	12 01 41.68	$+23 51 13.9$		"
Feb. 5.95729	11 46 27.98	$+28 17 16.6$		"
6.05016	11 46 14.84	$+28 20 51.2$		Vaghi
21.90583	11 01 25.32	$+37 46 34.8$	12.4	Waterfield

Cometary motion

In the absence of nongravitational effects and planetary perturbations, a small mass will orbit the sun on a path that is a conic section with the sun at one focus. The parameter that determines whether the path is an ellipse, a parabola, or a hyperbola is the total energy, E, of the body. The total energy, which is the sum of kinetic and gravitational potential energies, is given by

$$E = \frac{v^2}{2} - \frac{GM_\odot}{r} \tag{3.1}$$

Here v^2 is the velocity, r is the heliocentric distance, $M_\odot$ is the mass of the sun, G is the gravitational constant, and E is the total energy – which we express here as energy per unit mass. If one carries out the full derivation of the two-body problem, E can be shown to be a constant of motion. If the total energy of the body is negative, it is bound to the sun and will travel in an ellipse. If the total energy is positive, the body is unbound and will move in a hyperbola. The case of exactly zero total energy corresponds to a parabolic orbit. A body that falls toward the sun with initial zero velocity at infinity describes a parabola. Comets perturbed out of the Oort cloud by passing stars meet this criterion to a high approximation and should be observed to travel in nearby parabolic orbits.

The general equation for a conic section can be written in polar coordinates as

$$r = \frac{q(1 + e)}{1 + e \cos \theta} \tag{3.2}$$

where r is the distance from the focus that contains the sun, q is the perihelion distance, e is the eccentricity, and θ is measured from perihelion. The correspondence between orbital shape and total energy, described above, can be written in terms of the orbital eccentricity as shown in Table 3.1. The different types of orbits are illustrated in Figure 3.2.

Table 3.1. *Cometary orbit shapes*

Shape	Eccentricity	Energy
Ellipse	$0 \leq e < 1$	Negative
Parabola	$e = 1$	Zero
Hyperbola	$e > 1$	Positive

In the case of an elliptical orbit, equation 3.2 can be recast in the form

$$r = \frac{a(1 - e^2)}{1 + e \cos \theta} \tag{3.3}$$

where a is the semimajor axis. The perihelion distance q and the aphelion distance Q can be found from the expressions

$$q = a(1 - e) \qquad Q = a(1 + e) \tag{3.4}$$

Figure 3.2. Comparison of elliptic, parabolic, and hyperbolic orbits. The three orbits have the same perihelion and differ only in their eccentricity, e. The ellipse has $e = 0.9$, the parabola $e = 1.0$, and the hyperbola $e = 1.1$.

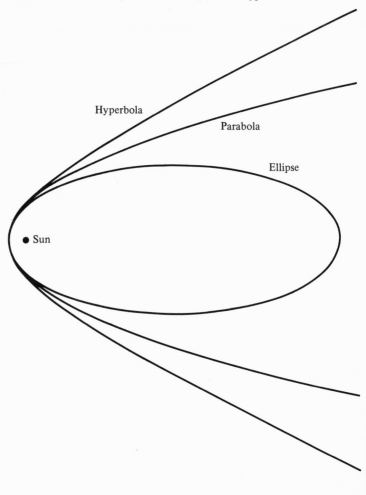

The theory of the two-body problem shows that, for elliptical orbits, the total energy per unit mass and the semimajor axis are related by the expression

$$E = - \frac{GM_\odot}{2} \cdot \frac{1}{a} \tag{3.5}$$

Equation 3.5 is also valid for parabolic orbits ($1/a = 0$) and hyperbolic orbits ($1/a < 0$). Thus, the velocity of the comet can be found by combining equations 3.1 and 3.5 to be

$$v^2 = GM_\odot \left\{ \frac{2}{r} - \frac{1}{a} \right\} \tag{3.6}$$

At perihelion,

$$v_p^2 = \frac{GM_\odot}{a} \cdot \frac{1 + e}{1 - e} \tag{3.7}$$

The analogous expression for the velocity for aphelion is easily found and can be combined with equation 3.7 to yield the simple relationship

$$v_p = v_a \cdot \frac{1 + e}{1 - e} \tag{3.8}$$

For a parabolic orbit $E = 0$, and therefore equation 3.1 gives the velocity at any point in the solar system

$$v^2 = \frac{2GM_\odot}{r} \tag{3.9}$$

This velocity can be interpreted as the velocity a body needs to escape from the solar system when it is at distance r from the sun. Conversely, it is the velocity at r of a body that has fallen from infinity. At the earth's orbit this escape speed or parabolic speed is 42 km sec^{-1}.

The general solution of the two-body problem can be characterized by six constants of integration that fully specify the size and shape of the orbit, its orientation in space, and the position of the body in the orbit. From the discussion above, it is clear that the total energy is one possible parameter. For astronomical applications, the orbital parameters are used to predict future positions of the body in the sky. Thus, the six parameters most frequently used are a set of independent *orbital elements* chosen for their convenience in calculating ephemerides. They are:

1. q, the perihelion distance in astronomical units,
2. e, the eccentricity,
3. T, the time of perihelion passage
4. i, the inclination of the orbital plane to the plane of the ecliptic

5. Ω, longitude of the ascending node (measured east from the vernal equinox)
6. ω, the angular distance of perihelion from the ascending node, also called the *argument of perihelion*

The elements of a cometary orbit are illustrated in Figure 3.3.

The quantities q and e are a measure of the size and shape of the orbit in the orbital plane; T fixes the comet in time along the path. The orbital plane's position in space is specified by i and Ω. The orientation of the orbit in the plane is fixed by ω. The quantities i, Ω, and ω are referred to a specific equinox, say 1950.0. The comet's orbit is direct if $0 < i < 90°$ and retrograde if $90° < i < 180°$. For an ellipse other quantities, such as the semimajor axis a and the period P, are easily derivable from geometrical relationships and Kepler's third law. The energy of an orbit and changes in energy are often described in terms of the reciprocal of the semimajor axis, $1/a$, which is positive for an ellipse, zero for a parabola, and negative for a hyperbola (see equation 3.5). Sample orbital elements are given in Table 3.2, along with a variety of computed quantities. Comets with nearly identical orbital elements (time of perihelion passage excepted) are members of comet groups; examples are sun-grazing comets 1843 I and 1880 I. Comet families consist of those comets with aphelia near the orbit of one of the Jovian planets. Several families were formerly recognized, but the earlier work did not fully consider that the orbital inclination would have to be close to zero

Figure 3.3. Schematic of a cometary orbit illustrating the elements as discussed in the text. The positions N_A and N_D mark the ascending and descending nodes, respectively. The ♈ is the astronomical symbol for the vernal equinox.

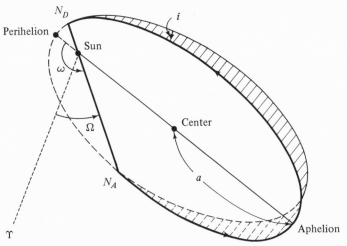

for many comets to be influenced by the outer planets. Only Jupiter's family is recognized today.

The simple geometrical shapes are strictly applicable only for the ideal case of a comet in orbit around a sun with no perturbing bodies such as planets. Accurate orbital calculations include planetary perturbations, and the results are expressed in terms of the so-called osculating elements that accurately reproduce the comet's position and velocity at a given instant in time, called the *epoch of osculation*. This time would be chosen, for example, to begin an accurate integration of an orbit with perturbations included. Osculating elements are a useful approximation when the perturbations cause only slow changes in the orbit. Obviously, they are of little value during a close planetary encounter.

Orbital determinations and ephemerides

In order to determine the six orbital parameters for a comet – or for any other body orbiting the sun – a minimum of three observations is required. Generally for a newly discovered comet, one initially assumes that the orbit is a parabola. The calculation of the elements then proceeds in a relatively straightforward fashion. After the comet has been observed many times over a period of a few months, a more accurate orbit at any epoch can be found; such an orbit is called a *definitive* orbit. The standard reference works on the techniques of orbital determination were written by Herget (1948), Dubyago (1961), and Escobal (1965).

Table 3.2. *Sample orbital elements*

Quantity	Comet Halley 1910 II	Comet Encke 1967 XIII
q	0.59 AU	0.34 AU
e	0.967	0.847
T	Apr. 20.2, 1910	Sept. 22.1, 1967
i	162°	12°
Ω	58°	334°
ω	112°	186°
P	76.1 yr	3.3 yr
a	17.96 AU	2.22 AU
v_p	54 km sec^{-1}	70 km sec^{-1}
v_a	0.9 km sec^{-1}	6 km sec^{-1}
v_p/v_a	60	12
Epoch	Apr. 20, 1910	Sept. 27, 1967

Given the six orbital elements of a comet, one can calculate its position in space. To find the position of the comet on the plane of the sky, one must know the earth's position in space as well. A table of geocentric positions of a comet is called an *ephemeris*. For bright or otherwise interesting comets, updated orbits and ephemerides are published regularly.

The theory of orbital perturbations of comets is not of immediate interest to us here. However, there is one result of the theory that is used in later discussion. During a cometary encounter with a planet, for example Jupiter, an approximation called the *Tisserand criterion* holds true. If the elements of a cometary orbit are determined before and after encounter, the Tisserand criterion is

$$\frac{a_j}{a_1} + 2 \left[\frac{a_1}{a_j} (1 - e_1^2)\right]^{1/2} \cos i_1' = \frac{a_j}{a_2} + 2 \left[\frac{a_2}{a_j} (1 - e_2^2)\right]^{1/2} \cos i_2'$$

(3.10)

Equation 3.10 is a special case of Jacobi's integral of energy and follows from the assumption that the orbital changes take place impulsively close to the planet. The Jacobi integral holds for a circular Jovian orbit around the sun, no other planets, and a comet of zero mass. In this equation, a_j is the semimajor axis of Jupiter's orbit, a is the comet's semimajor axis, e is the eccentricity of the comet's orbit, i' is the inclination to Jupiter's orbit, and the subscripts 1 and 2 refer to before and after encounter. In the past, equation 3.10 was used to test the identity of two comets appearing at two different epochs before laborious calculations were begun to follow the orbit in detail. At present, it is used to help clarify the physical processes involved in the origin of short-period comets.

We will dwell no further on the dynamics of any individual comet. The reader is referred to a number of standard textbooks on the subject. However, the statistics of the elements of the orbits of all well-studied comets give us considerable information on the origin of comets and comet families.

Cometary orbital statistics

The *Catalogue of Cometary Orbits* published by Marsden in 1979 provides material for statistical discussions of orbital elements. Figure 3.4 is a histogram showing the distribution of inclinations for parabolic and nearly parabolic comets.

The line on the figure indicates the expected distribution if the inclinations were random. The random distribution is proportional to $\sin i$, as can be seen if we consider the normals to the orbit. The normals to

Figure 3.4. The distribution of inclinations for parabolic and nearly parabolic comets. The solid line indicates the expected variation for a random distribution; see text for discussion. (Data from B. G. Marsden, 1979, *Catalog of Cometary Orbits*. Cambridge, Mass.: S.A.O. Central Bureau for Astronomical Telegrams)

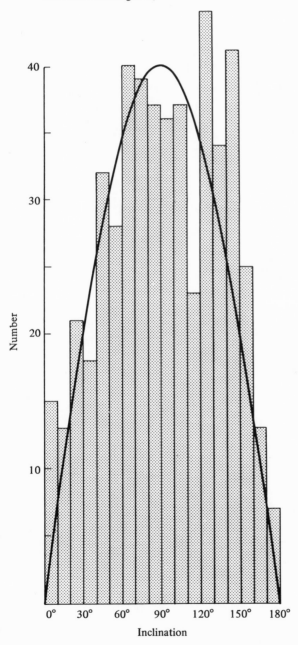

Figure 3.5. The distribution of inclinations for short-period comets. (Data from B. G. Marsden, 1979, *Catalog of Cometary Orbits*. Cambridge, Mass.: S.A.O. Central Bureau for Astronomical Telegrams)

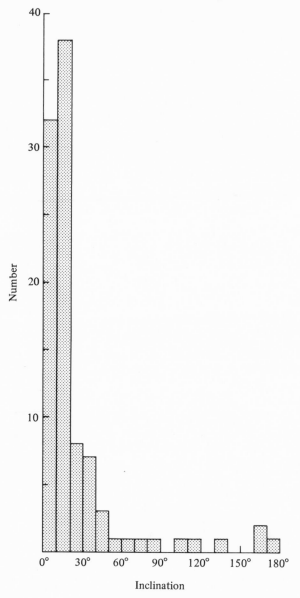

all orbits with inclinations between i and $i + di$ intersect the sky between the small circles with angular distances i and $i + di$ from the ecliptic pole. If the distribution of normals is random, the number of intersections is proportional to the area between the small circles, which is proportional to sin i. The distribution of inclinations for the long-period comets is approximately random, but there is a distinct relative excess of retrograde orbits.

The distribution of inclinations for short-period comets is illustrated in Figure 3.5. The differences between this distribution and that for the long-period comets is striking. Almost all short-period comets are in direct orbits with inclinations less than 30°.

The distribution of the periods for the short-period comets is shown in Figure 3.6. A huge peak between 7 and 8 years is the main feature. The majority of these comets have aphelia near the orbit of Jupiter (5.2 AU), as shown in Figure 3.7.

Selection effects as well as physical situations show up in the statistics of comet orbits. There is a strong correlation between q, a, and e, which are related for an elliptical orbit by $q = a (1 - e)$. For the long-period comets, we observe only those with $q \sim 1$ and, hence, a large value of a requires a value of e near 1. For the short-period comets such as Jupiter's family, a (strictly Q) is fixed, and hence there is a correlation between q and e.

The statistics of orbital parameters as illustrated in Figures 3.4 to 3.7 can be taken at face value to indicate a simple scheme for the origin of Jupiter's family of comets. The long-period comets, preferentially the direct ones, are depleted by encounters with Jupiter. This produces direct comets with low inclinations and aphelia near the orbit of Jupiter. The capture process could take place in one encounter, as illustrated in Figure 3.8. Specific examples of this process are known, such as Comet Brooks (1889 V), which encountered Jupiter and had its orbital period changed from approximately 29 years to approximately 7 years. (Of course, other results of such an encounter are possible.) However, the current view is that the short-period comets probably do not arise from single encounters with Jupiter. We will return to this subject when discussing the origin of comets.

Nongravitational forces

The final topic we will discuss relative to orbits and orbital parameters is that of nongravitational forces on comets (see Chapter 2). There is no longer any doubt concerning the reality of these forces, and instances of comets with increasing periods and comets with decreasing periods are known. The basic physical process is the loss of material (as suggested by Bessel in 1836) coupled with rotation of the cometary nucleus, as suggested by Whipple (1950a). There remains no reason to assign a substantive role to interactions with the interplanetary medium.

The configuration resulting in nongravitational acceleration is shown in Figure 2.8. For a nonrotating nucleus, the volatile ices sublimate primarily in the sunward direction and produce a force in the antisolar direction. In this case, the comet simply moves around the sun with a slightly reduced solar attraction. Kepler's third law, namely

$$\frac{P^2}{a^3} = \frac{4\pi^2}{GM_\odot} \tag{3.11}$$

Figure 3.6. The distribution of periods for short-period comets. (Data from B. G. Marsden, 1979, *Catalog of Cometary Orbits*. Cambridge, Mass.: S.A.O. Central Bureau for Astronomical Telegrams)

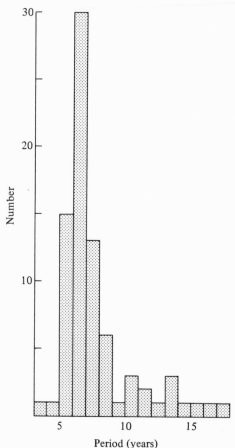

Figure 3.7. The distribution of aphelia for short-period comets. A few comets have aphelia larger than 14 AU. (Data from B. G. Marsden, 1979, *Catalog of Cometary Orbits*. Cambridge, Mass.: S.A.O. Central Bureau for Astronomical Telegrams)

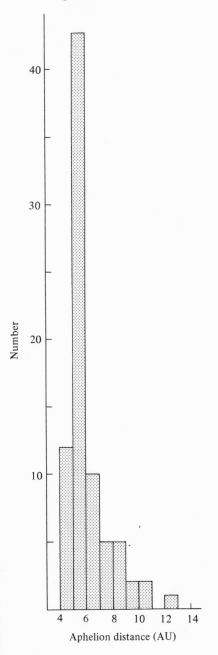

tells us that the effect of the reduced solar attraction will be an increase in the ratio P^2/a^3, which will last as long as significant sublimation occurs. However, if the nucleus is rotating and there is a time lag between maximum solar flux received and maximum mass loss, nonradial forces will exist. These forces will generally have components in the transverse direction and in the direction normal to the plane of the comet's orbit. The transverse component is responsible for the long-term changes in orbits. If the rotation causes mass loss in the forward direction with respect to the comet's motion, the force generated decelerates the comet, causes the comet to spiral in toward the sun, and decreases the orbital period. This case is shown in Figure 2.9. If the mass loss is in the backward direction with respect to the comet's motion, the comet is accelerated and spirals away from the sun, and the orbital period is increased. Note that a relatively slow rotation rate is necessary to produce detectable nongravitational forces. Very rapid rotation would smear out any preferential direction of mass loss.

Marsden and his associates have studied the nongravitational forces on comets extensively. Originally, the nongravitational forces were included empirically and the equation of motion was written as

Figure 3.8. Example of Jupiter perturbing a comet from a near-parabolic orbit into an elliptical orbit in one encounter. (After R. D. Chapman, 1974, *Space Science and Technology Today 1*: 1–40)

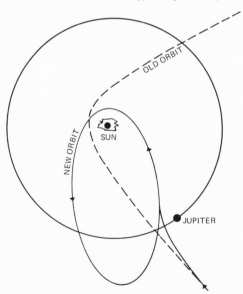

$$\frac{d^2\mathbf{r}}{dt^2} = \frac{-\mu\mathbf{r}}{r^3} + \frac{\partial R}{\partial\mathbf{r}} + A_1 \frac{e^{-B_1\tau}e^{-r^2/2}}{r^3}\hat{r}$$

$$+ A_2 \frac{e^{-B_2\tau}e^{-r^2/2}}{r^3}\hat{T} + A_3 \frac{e^{-B_3\tau}e^{-r^2/2}}{r^3}\hat{n} \qquad (3.12)$$

The symbols are as follows: The As and Bs are empirically determined constants; r is the radius vector; $\mu = GM_\odot$; R is the planetary disturbing function; τ is measured from the epoch of osculation in units of 10^4 ephemeris days; $\hat{r}$, $\hat{T}$, and $\hat{n}$ are unit vectors directed in the radial direction, in the orbital plane and 90° forward (in the sense of the comet's motion) of the radius vector, and in the direction normal to the orbital plane, respectively. The acceleration is given in astronomical units per (40 ephemeris days)^2.

Results from the earlier investigations can be summarized as follows. The radial nongravitational acceleration (in this discussion, when it can be accurately determined) is directed in the antisolar direction and is roughly an order of magnitude larger than the transverse nongravitational acceleration. This means that the lag angle is small. The transverse component is equally likely to be directed ahead of or behind the comet's motion. The nongravitational acceleration normal to the orbital plane is dynamically unimportant. The nongravitational forces seem to act continuously and may decrease slowly with time. In some cases, the nongravitational forces may have increased after a comet had a close approach to Jupiter.

Recent work has replaced the empirical nongravitational terms in equation 3.12 with terms based on the vaporization rate of water snow. Thus, Marsden, Sekanina, and Yeomans (1973) would write

$$\frac{d^2\mathbf{r}}{dt^2} = \frac{-\mu r}{r^3} + \frac{2R}{2\mathbf{r}} + A_1 g(r)\hat{r} + A_2 g(r)\hat{T} \qquad (3.13)$$

where

$$g(r) = \alpha \left(\frac{r}{r_0}\right)^{-m} \left\{1 + \left(\frac{r}{r_0}\right)^n\right\}^{-k} \qquad (3.14)$$

In equation 3.14, α is a normalization factor chosen so that $g(1) = 1$; r_0 is the distance beyond which the nongravitational forces (and presumably the vaporization of water snows) drop rapidly; and m, n, and k are constants. The vaporization flux can be written empirically as

$$A = Z_0 g(r) \qquad (3.15)$$

Data for vaporization of water snows and assumptions concerning the albedo led to the values (Table 3.3) as given by Delsemme and Delsemme (1971). Equation 3.14 is accurate to $\pm 5\%$ in representing the vaporization data. We will return to this discussion when Delsemme's model is presented, but we should mention that the total flux implied by equation 3.15 is compatible with values obtained by other means.

The procedure is then to solve for the constants A_1 and A_2 and thus determine the nongravitational forces. Basically, for periodic comets, it

Table 3.3. *Parameters for a vaporization controlled by water*

Parameter	Value
r_0	2.808 AU
Z_0	3×10^{17} molecules cm^{-2}-sec
m	2.15
n	5.093
k	4.6142
α	0.1113

Source: Delsemme and Delsemme (1971).

Figure 3.9. Variation with time of the transverse nongravitational parameter A_2 for Comet Encke. (Courtesy of B. G. Marsden and Z. Sekanina, Smithsonian Astrophysical Observatory)

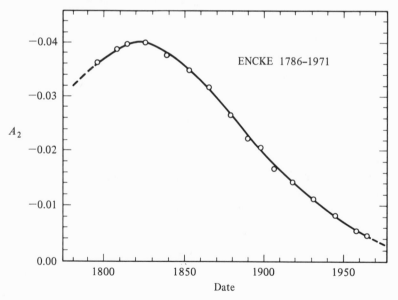

is the transverse term or A_2 that is determined by an orbit calculation. An example for the case of Comet Encke is shown in Figure 3.9. Note that A_2 is quite smooth and has been continuously decreasing since about 1825. Another result is that the nongravitational forces are important in calculating the so-called original values of orbital elements. For the long-period comets only A_1 is well determined, and it is always in the sense of being a radial force. An empirical result is that the radial force makes the orbit more elliptical; that is, it increases $(1/a)$ as compared to the purely gravitational case. This result means that the Oort cloud is closer to the sun than originally calculated.

Finally, Yeomans has discussed the nongravitational motion of Comet Kopff and reported that the transverse acceleration had changed direction (Figure 3.10). The phenomenon is also known for comets Pons–Winnecke and Faye. The simplest explanation in terms of the rotation of a water–ice nucleus is that the axis of rotation precesses. When $A_2 \approx 0$, the rotation axis would be nearly in the plane of the comet's orbit. Thus, changes in A_2, including changes in sign, can be produced without invoking exotic processes.

Figure 3.10. Variation with time of the transverse nongravitational parameter A_2 for Comet Kopff. (Courtesy of D. K. Yeomans, Jet Propulsion Laboratory)

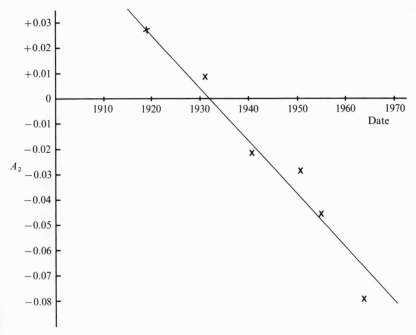

4 Structure of comets

The basic structural elements of comets have been discussed earlier (see Chapter 2). They are the nucleus, coma, tail, and hydrogen–hydroxyl cloud. To open this chapter, we will discuss briefly the observational techniques used to infer the properties of these structural components. To date, these techniques can be described as remote sensing methods. It is to be hoped that before the 1980s are over, these techniques will be supplemented by in situ probing of comets.

Observing techniques

Prior to approximately 1881 (see Chapter 2), cometary observations were carried out visually. An experienced observer with a good telescope (then usually a refractor) could measure accurate positions and observe structure near the nucleus. Some of the drawings made by visual observers show detail that cannot be fully captured even today by photography through large telescopes. Naked-eye observations or observations through defocused binoculars are useful in determining the total magnitude of the cometary head. Similar observations over a period of time lead to the cometary light curve. Experienced observers can produce a light curve that is internally consistent to within a few tenths of a magnitude, but comparison of light curves made by different observers often reveals extraordinary differences in the magnitude level of the curves. No seemingly straightforward observational quantity is (and should be) regarded with as much suspicion as cometary magnitudes.

Frequently one can distinguish between the nuclear magnitude m_1 and the total magnitude m_2 of the cometary head. The nuclear magnitude refers to the brightness of the central, quasi-stellar concentration of light or photometric nucleus (sometimes called the *false nucleus*). This concentration of light may or may not be due to the comet's physical nucleus, but if it is, the nuclear magnitude contains information about the size of the nucleus. The total magnitude refers to the entire material in the head as well as its rate of production. Again, the difficulties with magnitudes should be noted; the magnitude derived can depend on the size of the telescope used, contrast effects, and so on. Nuclear magnitudes are typically two to five magnitudes larger (fainter) than total magnitudes.

Photography remains a prime source of data on comets. Photographs can vary from short, astrometric exposures designed to produce a small image suitable for positional determination to long, wide-angle exposures designed to show the extent of and structure in comet tails. Color emulsions can be used as well as the standard emulsions available on film or plates. Filter–emulsion combinations can be used to isolate an interesting wavelength region or the emission from a particular molecule. For example, Kodak IIa-0 plates with a UG-1 filter produce excellent photographs of plasma tails that emit strongly in the blue due to their principal constituent, CO^+. Photographs normally produced are not suitable for studies of the nucleus or detailed structure near the nucleus.

Photoelectric photometry also has found a place in cometary research. However, the usual wide-band photometry, such as the UBV system, has not been particularly useful to date. The typical method of cometary photometry is to use a pair of filters, one of which isolates emission from a particular species and the other of which isolates a nearby continuum, often selected to be as free as possible of emission lines. For example, photoelectric photometry of the CN band at 3880 Å could be carried out as follows: The cometary brightness can be measured alternatingly through an interference filter or intermediate passband filter (full width at half transmission of order 75 Å) having a maximum transmission near 3880 Å and through a filter similar to the U filter of the UBV system (full width at half transmission around 700 Å, maximum transmission around 3600 Å). The CN emission relative to the continuum can be derived from a color calculated in the usual way by

$$U - E = -2.5 \log \frac{F_u}{F_e} + \text{const} \qquad (4.1)$$

where F_u and F_e are the measured fluxes in the U color and in the narrow passband near 3880 Å denoted by E, respectively. These colors can be used to define a color difference,

$$D = (U - E)_{\text{comet}} - (U - E)_s \qquad (4.2)$$

between the comet and a source s (e.g., stars or the sun) with a continuum distribution similar to the comet under study. When $D = 0$, no CN emission is present, and the value of D increases with increasing strength up to a maximum value depending on the details of the filter transmission curves.

The general scheme outlined above can be adapted to the study of any cometary emission line or band. Note that the projected size, at the comet, of the entrance diaphragm used in the photometry should be specified for the measurements to be useful.

Figure 4.1. Photoelectric scans of two comets. (*a*) Comet Arend–Roland.
A yellow filter was used to eliminate effects of overlapping orders for
wavelengths greater than 5000 Å. (*b*) Comet Mrkos. In the long-wavelength
region, the cometary scan (dashed line) is similar to the scan of Capella (dotted
line), a yellow star approximately the same color as the sun. (Courtesy of W.
Liller, Harvard University)

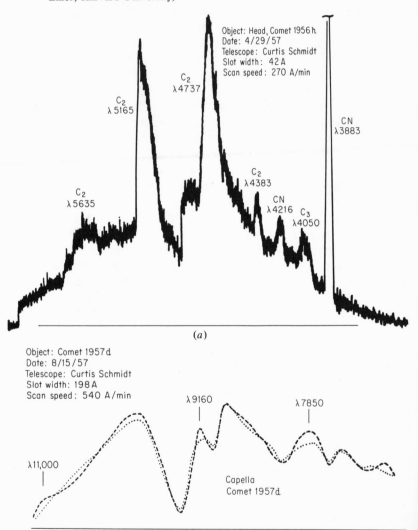

Object: Head, Comet 1956 h
Date: 4/29/57
Telescope: Curtis Schmidt
Slot width: 42 A
Scan speed: 270 A/min

C_2
$\lambda 4737$

C_2
$\lambda 5165$

CN
$\lambda 3883$

C_2
$\lambda 4383$

C_2
$\lambda 5635$

CN
$\lambda 4216$ C_3
$\lambda 4050$

(*a*)

Object: Comet 1957 d
Date: 8/15/57
Telescope: Curtis Schmidt
Slot width: 198 A
Scan speed: 540 A/min

$\lambda 9160$

$\lambda 7850$

$\lambda 11,000$

Capella
Comet 1957 d

(*b*)

Cometary spectroscopy can be carried out usefully at a variety of dispersions ranging from roughly 500 Å mm^{-1} with objective prisms to high dispersions of approximately 1 Å mm^{-1}. The spectra have been recorded both photographically and photoelectrically. Photoelectric spectrum scans are illustrated in Figure 4.1. Note the large variations in the appearance of the spectra. Examples of low-dispersion spectrograms are shown in Figure 4.2. High-dispersion spectra of the CN bands near 3880 Å are shown in Figures 4.3 and 4.4. Additional examples of slit spectrograms are shown in Figures 4.5, 4.6, and 4.7. A summary of atoms and molecules observed in comets is given in Table 4.1.

Analysis of the spectroscopic data provides several important pieces of information, including: (1) identification of atomic and molecular species, (2) abundances of these species, (3) bulk motions in comets from the Doppler effect, and (4) excitation processes in the cometary

Figure 4.2. Low-dispersion spectrograms of comets Arend–Roland and Mrkos in visual wavelengths. (Lick Observatory photograph)

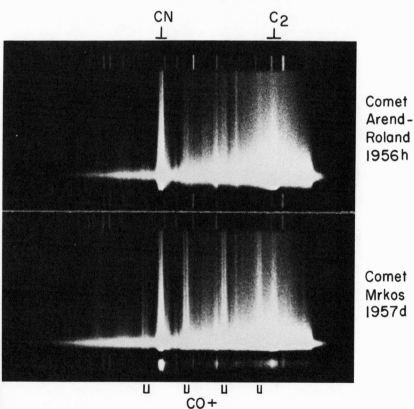

atmosphere. Each of the enumerated points is useful individually. However, cometary spectroscopic analysis has proceeded to the point where attempts are made to find an internally consistent set of all the parameters. In this case, synthetic spectra are computed, with all the known, relevant physical processes included. An acceptable synthetic spectrum must fit the observed spectrum in detail.

Radio wavelength studies of comets have become a major source of

Figure 4.3. High-dispersion spectrum of the CN bands near 3880 Å in Comet Mrkos. Irregularities in the CN intensities caused by the (marked) solar absorption lines are clearly shown. (Courtesy of J. L. Greenstein, Hale Observatories, California Institute of Technology)

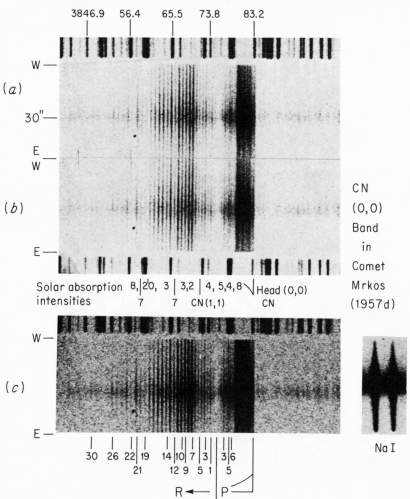

information since about 1970. Earlier claims of detection in radio wavelengths are either spurious or not useful. Recently, though, the molecules CH_3CN (methyl cyanide), HCN (hydrogen cyanide), CH, and OH have been detected at millimeter and microwave wavelengths, though with small signal-to-noise ratios. Finally, enhanced interplanetary scintillations have been detected and observed when a comet passed in front of a celestial radio source. Water vapor may have been detected in Comet Bradfield (1974 III).

Infrared observations in the range 1–20 μm with relatively coarse spectral resolution have been carried out for some time (see Figures 7.8 and 7.23) and are sufficient to establish the intensity of the continuous emission. From these data one can calculate the temperature of the

Figure 4.4. Spectrograms of the CN bands near 3880 Å in Comet Bennett, showing resolution of the band head. (*a*) Observatoire de Haute Provence spectrogram at original dispersion of 7 Å mm^{-1}. (*b*) Hale Observatory spectrogram (taken by G. W. Preston) at original dispersion of 4.5 Å mm^{-1}. (Courtesy of C. Arpigny, Institut d'Astrophysique, Liège)

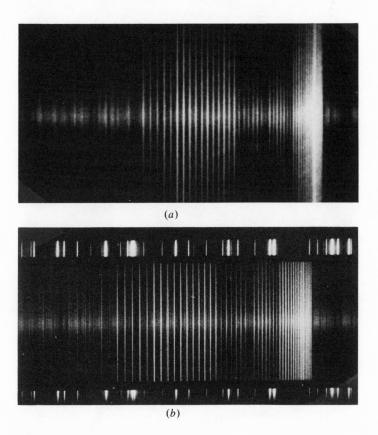

(*a*)

(*b*)

Figure 4.5. Spectrograms of the head region of comets. All wavelengths in Angstroms. (*a*) Eclipse comet (1948 XI). McDonald Observatory spectrogram taken by P. D. Jose and P. Swings at original dispersions of 50 Å mm⁻¹ at 4000 Å, and 200 Å mm⁻¹ at 5500 Å. (*b* and *c*) Comet Tago–Sato–Kosaka (1969 IX). Observatoire de Haute Provence spectrograms taken at original dispersion of 20 Å mm⁻¹. (*d*) Comet Bennett (1970 II). Observatoire de Haute Provence spectrogram taken at original dispersion of 13 Å mm⁻¹. (Courtesy of C. Arpigny, Institut d'Astrophysique, Liège)

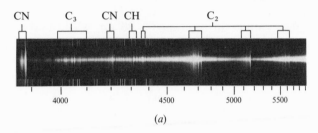

(*a*)

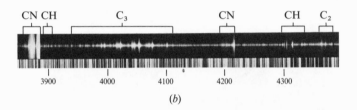

(*b*)

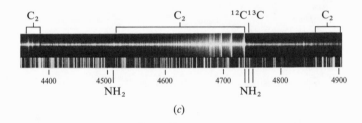

(*c*)

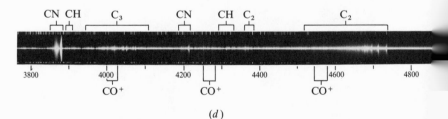

(*d*)

grains and gain some insight into their composition, for example, by observation of the silicate signature at 10 and 18 μm in Comet Bennett (1969i), as reported by Maas, Ney, and Woolf (1970). The 1–20 μm region is also very important because the vibration–rotation transitions of many molecules fall in it. A sample of the possibilities is listed in Table 4.2.

We may have some additional information on the nature of cometary dust particles as well. Since 1974, extraterrestrial particles thought to be cometary in origin have been collected in the atmosphere at altitudes of 20 km by U-2 aircraft. When pure aluminum oxide (Al_2O_3) spherical particles generated by solid fuel rocket engines are excluded, approximately half the collected particles have elemental abundances that closely match the abundances of carbonaceous chondrite meteorites. Scanning electron microscope pictures of two of these particles are shown in Figure 4.8. Notice the porous nature of the structures. This class of particles, according to Brownlee, Rajan, and Tomandl (1977:134), "seems to come from the gentle fragmentation of a single type of parent-body material: a black aggregate of grains mainly 1000 Å

Table 4.1. *Atoms, ions, molecules, and molecular ions observed in comets*[a]

Head	Tail
H, OH, H_2O, O, S C, C_2, C_3, CH, CN, CO, CS HCN, CH_3CN, NH, NH_2, Na,Fe, K, Ca, V, Cr, Mn, Co, Ni, Cu	CO^+, CO_2^+, H_2O^+, OH^+, CH^+, N_2^+, Ca^+, C^+, CN^+

[a] Isotopes not indicated.

Figure 4.6. Spectrogram of the tail region. Comet Bennett (1970 II). Observatoire de Haute Provence spectrogram taken at original dispersion of 20 Å mm^{-1}. Wavelengths in Angstroms. (Courtesy of C. Arpigny, Institut d'Astrophysique, Liège)

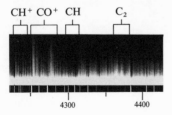

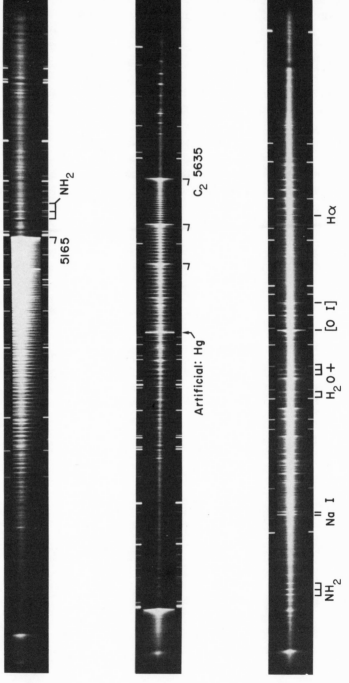

Figure 4.7. High-dispersion spectrograms of Comet Kohoutek (1973f) on January 9 and 11, 1974. (Lick Observatory photograph)

in size.'' The particles do not appear altered or strongly heated by entry into the atmosphere.

The extraterrestrial origin of these particles seems definite because helium atoms implanted by the solar wind have been detected. Although these particles probably come from comets, this conclusion is still tentative.

Finally, recent years have seen the observation of comets from above the atmosphere (Chapter 7). Photometry of ultraviolet emission lines, particularly $\lambda 1216$ Å, the Lyman α line of neutral hydrogen, has been carried out on several comets from satellites in earth orbit; ultraviolet lines of other constituents such as OH have also been observed. Rockets have been used to obtain ultraviolet images and spectrograms. Observations of OH bands near 3100 Å have been made from aircraft.

Comet Kohoutek was observed extensively by the astronaut crew of *Skylab IV* during December 1973 and January 1974. Instrumentation on board *Skylab* was used to obtain photometry of the Lyman α halo. In addition, scientist–astronaut E. G. Gibson made visual observations of the comet near perihelion. These observations showed the development of Comet Kohoutek's antitail. Sketches showing the spectator's view from *Skylab* are shown in Figure 4.9. The *Skylab* observations will be discussed later.

Tails

Photographs of well-developed comets such as Comet Mrkos (Figure 4.10) generally show two distinct types of tails, called *type I* and *type II* tails. Type I tails are straight and show considerable fine structure. Type II tails are curved and relatively lacking in fine structure. In a color photograph, type I tails appear blue and type II tails appear yellow. Occasionally one finds reference to type III tails. These are

Table 4.2. *Some interesting infrared molecular emissions expected from comets (H_2O excluded)*

λ (μm)	Molecule	λ (μm)	Molecule
2.68	OH	5.25	NO
3.03	NH	5.48	CN
3.2	CH_4	3, 6, 8, 14	NH_3
4.26	CO_2	7.5	CH_4
4.61	CO	14.0	C_2H_2
4.76	CH	15.05	CO_2

Figure 4.8. Scanning electron microscope pictures of possible cometary dust particles collected in the earth's atmosphere. (Courtesy of D. Brownlee, University of Washington)

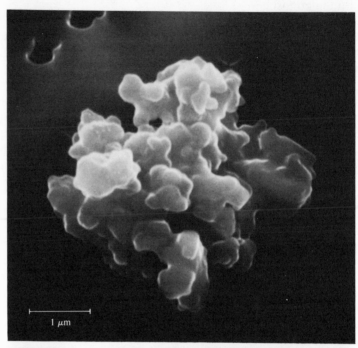

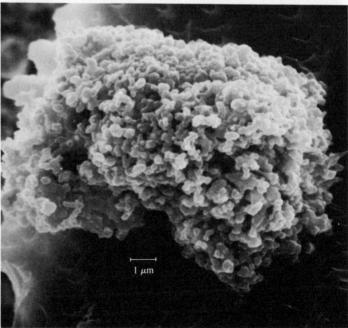

Figure 4.9. Sketches of Comet Kohoutek near perihelion as seen from *Skylab*. (Courtesy of E. Gibson, NASA–Johnson Space Center)

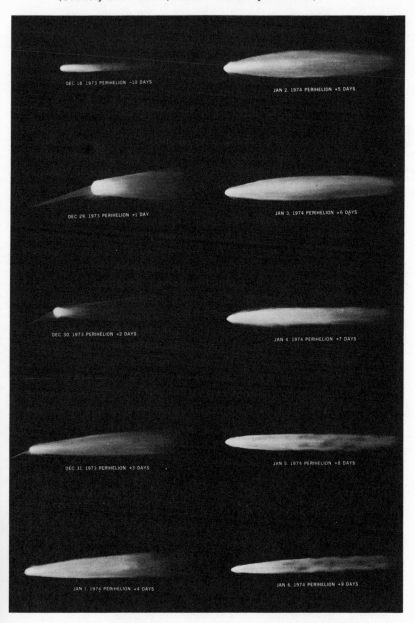

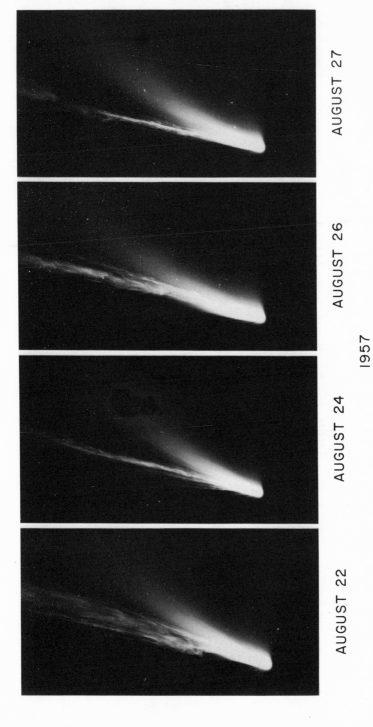

AUGUST 27

AUGUST 26

1957

AUGUST 24

AUGUST 22

Figure 4.10. Comet Mrkos as photographed with the 48-in Schmidt telescope. (Hale Observatories)

88

strongly curved tails. Types II and III tails are composed of dust, and their color is due to reflected sunlight.

Spectroscopic studies of type I tails show that they consist of ionized molecules. Thus, type I tails are also referred to as *plasma tails*. The major emitter in the visual range is the CO^+ ion, which is responsible for the observed blue color of the tail. Emissions due to the molecular ions CO_2^+, H_2O^+, OH^+, CH^+, and N_2^+ have been found in plasma tails. Densities of CO^+ inferred from total intensity measurements in the molecular bands range from around 10^3 cm^{-3} near the head to 10 cm^{-3} far from the head.

Plasma tails are made up of a number of thin bundles of material called *tail rays* or *streamers*, with radii in the range 2000–4000 km. Total widths of typical plasma tails are in the range 10^5–10^6 km and their lengths reach and occasionally exceed 10^8 km or 1 AU. The tails make an angle of a few degrees with the prolonged radius vector from the sun, with the tail lagging behind the comet's motion.

Tail streamers (Figure 4.11) are compelling evidence for the importance of magnetic fields in comets. For any reasonable value of the kinetic temperature in comet tails, a CO^+ ion will move a distance

Figure 4.11. Head of Halley's comet on May 8, 1910. The tail streamers are clearly shown. (Hale Observatories)

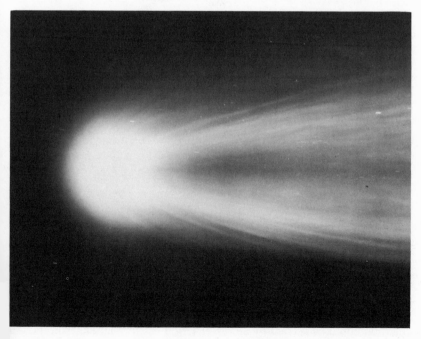

comparable to the diameter of a streamer in a few hours. Thus, thermal motions should wash out any fine structure. There are two situations in which thermal motions will not destroy fine structure: (1) if the kinetic temperature is in the range $0 < T_K < 1$ K or (2) if magnetic fields are present. Situation 1 is clearly unreasonable. However, the presence of fine structure can be easily explained if one hypothesizes the presence of a magnetic field with field lines running along the tail. The CO^+ ions will move along a helical path, with a radius (the Larmor radius), r_L, given by

$$r_L = \frac{v_\perp m}{qB} \qquad (4.3)$$

In equation 4.3, $v_\perp$ is the component of the ion velocity perpendicular to the field lines, m is the ionic mass, q is the charge in electromagnetic units and B is the magnetic field in gauss. If we assume a $v_\perp$ corresponding to the thermal speed for 10^4 K and $B \approx 5 \times 10^{-5}$ G $= 5\gamma$ (a value comparable to the quiet solar wind field), we find $r_L \sim 10^2$ km. Any reasonable set of parameters will have an r_L less than the observed streamer sizes.

Type I tails also show kinks and knots of material. Generally, these structural details show motion and acceleration away from the head. Typical velocities range from 10 km sec^{-1} near the head to 250 km sec^{-1} far from the head. Accelerations are expressed in terms of the quantity $(1 - \mu)$, which is the net outward force acting on the structures, expressed in units of the local solar gravity. The typical values are $(1 - \mu) \approx 100$, but with wide variations. The values found in the outer rays near the head are larger than the values found in the inner part of the tail; there is apparently some kind of shielding effect. Currently, the tendency is to interpret the motions and accelerations of structural details as real physical motions. However, the evidence is not entirely convincing, and other possibilities such as wave motion should be considered. For example, the structures could be waves moving at the Alfvén speed V_A in the cometary plasma. The Alfvén speed in centimeters per second is given by

$$V_A = \frac{B}{(4\pi\rho)^{1/2}} \qquad (4.4)$$

where B is the magnetic field in gauss and ρ is the density in gm cm^{-3}. If the magnetic field is roughly constant in the main part of the tail, as argued just below, then a variation in density from high values near the nucleus to low values far from the nucleus could produce the observed speeds and apparent accelerations of features in the plasma tails. The question could be settled by an observation of Doppler shifts in the tail with favorable geometrical circumstances or by in situ measurement. The resolution of this basic question may take some time to achieve.

The origin of the CO⁺ plasma and the evolution of the tail streamers have been studied observationally. Some of the features are shown in Figures 4.12, 4.13, and schematically in 4.14. The CO⁺ plasma originates in a small volume near the sunward side of the nucleus. The region in which ionization occurs is confined to within 10^3 km of the nucleus. Jets of emission appear that initially point sunward but then bend back toward the tail. The jets have lifetimes of about 1 day, and during this time they lengthen and turn toward the tail axis, as illustrated in Figures 4.13 and 4.14. The process has been likened to the picture of a folding umbrella. Thus, the structure of the CO⁺ emission

Figure 4.12. Comet Morehouse in 1908. The same photograph is printed three times to bring out greater detail near the nucleus. (From *Atlas of Cometary Forms,* by J. Rahe, B. Donn, and K. Wurm, NASA SP-198, Washington, D.C.: GPO, 1969)

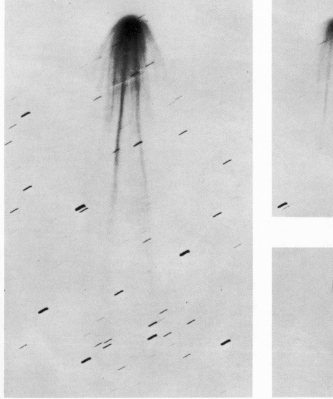
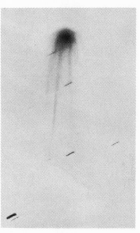

Figure 4.13. Comet Kobayashi–Berger–Milon. Sequence of photographs showing the turning of tail rays (or streamers) to the tail axis. (Joint Observatory for Cometary Research, NASA–Goddard Space Flight Center and New Mexico Institute of Mining and Technology)

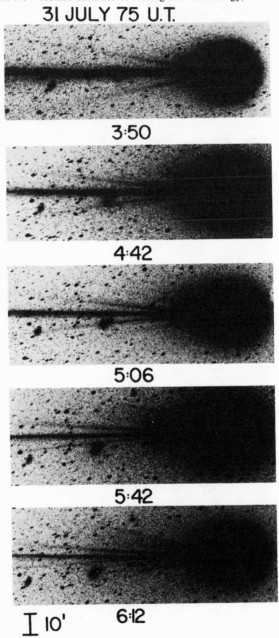

31 JULY 75 U.T.

3:50

4:42

5:06

5:42

I 10'   6:12

(and, by implication, the magnetic field) near the nucleus is complex, but away from the head the structural features are basically linear and oriented along the tail axis. In discussing waves and plasma instabilities in the tail, a useful first approximation is to consider the magnetic field lines as oriented parallel to the tail axis and to have uniform spacing within a cylinder with dimensions roughly equal to the observed extent of the tail emission.

At present there is no reason to doubt that the CO^+ originates from neutral CO (or perhaps CO_2) that is released from the nucleus or produced near the nucleus by the dissociation of more complex molecules. Hence, the ionization mechanism becomes a perplexing question. The speed of the expanding neutral gas in the coma is found to be approximately 0.5 km sec^{-1}. The travel time across the 10^3-km region in which ionization occurs implies an extremely short time scale for ionization, that is, $\sim 10^3$–10^4 sec. These times can be compared with the estimated time scale for photoionization of CO, which is $\sim 10^6$ sec. Recent ideas involving electrical discharges discussed in Part III may provide the answer.

The ion tails are normally found only in comets with heliocentric distances less than 1.5–2.0 AU and, for a time, it was thought that this restriction might be a clue to the ionization mechanism. Even here there are problems because of unmistakable exceptions to the general rule. Comet Humason (1961e) showed strong CO^+ emission at heliocentric distances of 5 AU. At the same time, the comet displayed unusual forms (Figure 4.15) that had an extremely turbulent or disrupted appearance.

Type II tails are generally not as long as type I tails, but they can still reach lengths of 10^7 km. Observations made when the earth is near the orbital plane of comets indicate that the dust tails are essentially flat structures. The spectrum of type II tails is basically a reflected solar spectrum, and hence the conclusion is that these tails are composed of dust. Typical dimensions of order 1 μm for the dust particles are inferred from polarimetry and photometry of the dust tails. The 1-μm size

Figure 4.14. Schematic drawing showing the lengthening and turning of tail streamers.

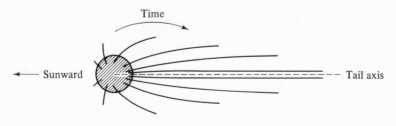

for the dust particles is also compatible with the observed curvatures of type II tails (see below). Type II tails also differ from type I tails in that they are homogeneous and somewhat featureless. Exceptions are (1) the synchronic band structure discussed below and (2) the fact that the boundary or edge of the dust tail is sharper on the convex side (near the plasma tail) than on the concave side (away from the plasma tail).

A few comets have been observed to exhibit additional antitails apparently pointing in the sunward direction. An antitail was observed on Comet Kohoutek (see Chapter 7), but the most famous one was observed on Comet Arend–Roland (1957 III). The latter comet exhibited a sunward fan (Figure 4.16) as the earth approached the plane of its orbit, and as the earth passed through the plane, a "sunward spike" (Figure 4.17) appeared. We believe that these features are sunward only in perspective, and a possible geometry is shown in Figure 4.18. The dust particles responsible for the appearance of this anomalous tail are probably much larger than the dust found in type II tails. As such the grains experience a small or negligible repulsive force, and instead of being blown out of the solar system they would be dispersed along or near the orbit of the comet. The dust particles probably constitute the particles that made up meteor streams.

Type II or dust tails appear to be basically dust released from the nucleus and acted upon by solar radiation pressure. The current theory

Figure 4.15. Comet Humason on August 6, 1962, as photographed by R. Rudnicki and C. Kearns, who noted the corkscrewlike structure. The comet was more than 2.5 AU from the sun. (Hale Observatories)

resembles the historical Bessel–Bredichin theory of tail forms discussed in Chapter 2, and it is – but with significant modifications. In the Bessel–Bredichin theory, when dust particles leave the nuclear region, each particle travels on a trajectory determined solely by solar gravity and solar radiation pressure. The particle orbits are not the observed form of the tail. The tail shape is the curve describing the location at a given instant of particles previously emitted from the head (see Figure 2.2).

Two limiting cases of the Bessel–Bredichin theory are of interest. The first case is the continuous expulsion of particles. One can define a curve called a *syndyname* (or *syndyne*), which is the instantaneous curve on which one finds continuously emitted particles with a given value of the solar repulsive force, or a given particle size or a given value of $(1 - \mu)$, depending on which parameter one chooses to specify. Syndynes are curves tangent to the radius vector at the com-

Figure 4.16. Comet Arend–Roland on April 22, 1957 showing the sunward fan. (Courtesy of R. Fogelquist, Bifrost Observatory)

et's head, and with a curvature away from the direction of the comet's motion. If dust particles of a fixed size, or a small range of sizes, were expelled from the nucleus with negligible initial velocity, the dust tail would look like a syndyne. If the dust were expelled in all directions with significant velocity, the tail would be broadened. If the dust particles had a significant range of sizes, then the tail would be made up of a series of syndynes and would appear broad.

Figure 4.17. Comet Arend–Roland on April 25, 1957 showing the sunward spike. (Lick Observatory photograph)

The second limiting case of interest is the instantaneous emission of particles. One can define a curve called a *synchrone,* which is the position of a group of particles ejected at the same instant but having a broad range of sizes or values of $1 - \mu$. A synchrone has a rectilinear form and makes an angle with the radial direction that increases with time. Figure 4.19 shows examples of syndynes and synchrones calculated for Comet Arend–Roland. Until the 1950s it was felt that the shapes of type II comet tails were explainable in terms of syndynes.

Pure type II comet tails (i.e., comets with negligible plasma tails) are rare, but a number of examples are known. Osterbrock studied comets Baade (1954h) and Haro–Chavira (1954k) and found that their tails lagged the radius vector by ≈ 45°. Because these comets were observed at distances between 4 and 5 AU and because other comets observed closer to the sun had radial type II tails near the head, this atypical orientation was ascribed only to comets observed far from the sun. However, Belton (1965) soon found that the ≈ 45° orientation was normal for all pure dust tails regardless of heliocentric distance. The presence of the plasma tail in mixed comets appears to account for the near-radial orientation of the dust tail through a drag force exerted on the dust particles by the streaming plasma. In the case of Comet Mrkos, the two types of tails are apparently decoupled within 10^6 km of

Figure 4.18. Diagram showing relative positions of earth and Comet Arend–Roland in April and May, 1957. On April 22.9 the earth is still somewhat away from the comet's orbital plane, and the particles producing the anomalous tail appear as a fan-shaped appendage (see Figure 4.16). The earth crossed the orbital plane of the comet on April 25.7. Thus, on April 25.3 the earth was quite close to the plane, and the anomalous particles appeared as a spike (see Figure 4.17). By May 4.5 the geometrical circumstances had changed so much that the anomalous particles were not viewed as an apparently sunward appendage. (Adapted from a figure by N. B. Richter)

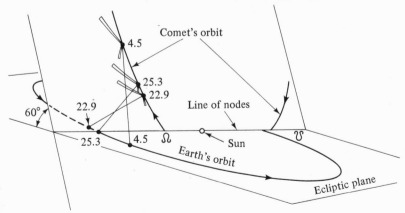

the nucleus. Thus, syndynes with their near-radial orientations near the head cannot explain the pure type II tail orientations. Synchrones can be constructed to fit some orientations, but not all of the orientations. Hence, the Bessel–Bredichin theory was thought to be inapplicable. Several alternate ideas (e.g., drag forces, charged dust particles) were tried but without success.

The solution to the problem was obtained by Finson and Probstein in 1968. Their approach, in retrospect, was fairly simple; they dropped the limiting forms of the Bessel–Bredichin theory and allowed particles of all sizes to be emitted at a rate that varied with time. We assume that the ejection speed v_i is given by conditions in the expanding coma and its interaction with the dust particles. Finson and Probstein found that the superposition of synchrones broadened by the initial speed and having a varying dust-ejection rate can produce an excellent fit to the observations, as shown in Figure 4.20.

The calculations can be carried out by utilizing either of two alternate approaches: (1) calculate the surface density of dust in the tail for one particle size and then integrate over all relevant particle sizes (i.e., a differential syndyne approach); (2) calculate the surface density of dust in the tail for all particles emitted at one time and then integrate over all

Figure 4.19. Syndynes (solid curves) and synchrones (dashed curves) for Comet Arend–Roland on April 27.8, 1957. The coordinate ξ is in the radial direction and η is perpendicular to ξ and directed opposite to the comet's orbital motion (i.e., in the orbital plane). Dates listed refer to the time of emission for the synchrones. (Courtesy of M. L. Finson and R. Probstein, Massachusetts Institute of Technology)

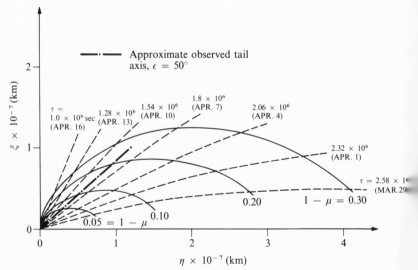

the times of emission that contribute to the tail (i.e., a differential synchrone approach). Here we describe the differential syndyne approach.

The desired quantity is the surface density of dust particles in the tail times the efficiency for light scattering. We begin by deriving a simple relation between particle area and $(1 - \mu)$, which is defined by

$$1 - \mu = \frac{F \text{ (radiation)}}{F \text{ (gravity)}} \tag{4.5}$$

Figure 4.20. Calculated and measured isophotes for Comet Arend–Roland on May 2.9, 1957. The M coordinate is in the apparent radial direction, with the N coordinate perpendicular to M and directed opposite the comet's orbital motion. (Courtesy of M. L. Finson and R. Probstein, Massachusetts Institute of Technology)

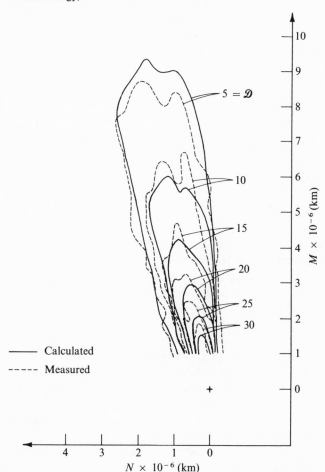

The radiation pressure is given by standard expressions and $F_{rad} \propto Q_{PR}a^2$ where Q_{PR} is the scattering efficiency for radiation pressure and a is the radius of the (assumed) spherical particles. In an obvious way, $F_{grav} \propto \rho_d a^3$, where ρ_d is the density of the dust particles. Hence, we can write

$$1 - \mu = \frac{D}{\rho_d a} \tag{4.6}$$

where

$$D = 0.6 \times 10^{-4} \, Q_{PR} \, (\text{g cm}^{-2}) \tag{4.7}$$

The amount of scattered light is proportional to the surface area $(\rho_d a)^2 \, g(\rho_d a)$, where $g(\rho_d a)$ is the distribution function for sizes as a function of $\rho_d a$. It is more convenient to work with $(1 - \mu)$ as a variable, and the change is easily made, considering that $g(\rho_d a)$ is the number of particles per unit interval in $(\rho_d a)$. Changing the independent variable from $\rho_d a$ to $1 - \mu$, with the aid of equation 4.6 we find

$$(\rho_d a)^2 \, g(\rho_d a) \, d(\rho_d a) \propto f(1 - \mu) \, d(1 - \mu) \tag{4.8}$$

where $f(1 - \mu)$ is the distribution function for particle sizes as a function of $(1 - \mu)$. A relatively simple derivation shows that surface density of dust (in terms of light-scattering efficiency) in a differential syndyne tail is proportional to

$$\dot{N}_d f(1 - \mu) \, d(1 - \mu) \left[2v_i \tau \frac{dx}{d\tau} \right]^{-1} \tag{4.9}$$

Here $\dot{N}_d$ is the rate of dust emission, v_i is the initial injection speed, τ is the time since emission (variable along the syndyne), and x is the distance along the axis of the syndyne. Physically, $\dot{N}_d f(1 - \mu) \, d(1 - \mu)$ gives the effective surface density of scattering particles at a point. The terms $2v_i \tau$ and $dx/d\tau$ give the reduction in surface density due to the dispersion in the lateral and longitudinal directions, respectively, with respect to the tail axis.

The integrations over $(1 - \mu)$ according to equation 4.9 require the specification of three functions, namely, $\dot{N}_d(t)$, $f(1 - \mu)$, and $v_i(1 - \mu, t)$. These functions were assumed initially and adjusted to produce the excellent agreement for Comet Arend–Roland shown in Figures 4.20 and 4.21. Numerical experiments showed that the solutions were essentially unique. The results for $\dot{N}_d(t)$ and $f(1 - \mu)$ are shown in Figures 4.22 and 4.23. Note the peaked distribution in the dust emission before perihelion $(t = 0)$. This result is entirely in accord with the observation that comets approaching the sun for the first time are generally dusty and tend to be dustier before perihelion. The size distribution can be expressed in terms of ($\rho_d a$), as shown in Figure 4.24. For a density $\rho_d = 3$ g cm^{-3}, the peak in the particle distribution is for dust with particle diameters ≈ 1 μm. The information concerning the size distribution of particles shown in Figure 4.24 is entirely consistent with the photometric and polarimetric data on the dust in type II comet tails.

The absolute value of the scattered light intensity in the tail determines the rate of dust emission. Typical rates found are in the range

Figure 4.21. Calculated and measured isophotes for Comet Arend–Roland on April 27.8, 1957, showing the ability to reproduce detail. The calculations were not extended to reproduce the spike, although in principle this could have been done. (Courtesy of M. L. Finson and R. Probstein, Massachusetts Institute of Technology)

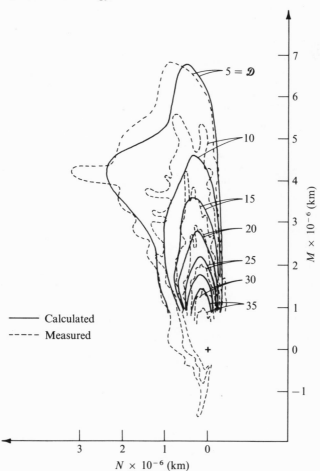

Figure 4.22. Variation of the relative dust-particle emission rate $\dot{N}_d$ with time. (Courtesy of M. L. Finson and R. Probstein, Massachusetts Institute of Technology)

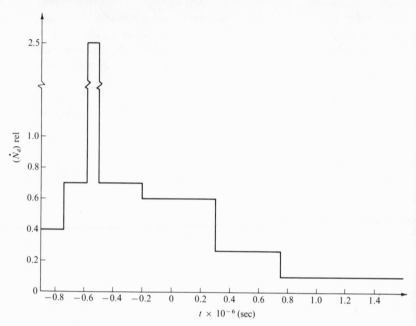

Figure 4.23. The particle-size distribution function $f(1 - \mu)$. The dashed line is the distribution used to represent the outburst shown in Figure 4.22. (Courtesy of M. L. Finson and R. Probstein, Massachusetts Institute of Technology)

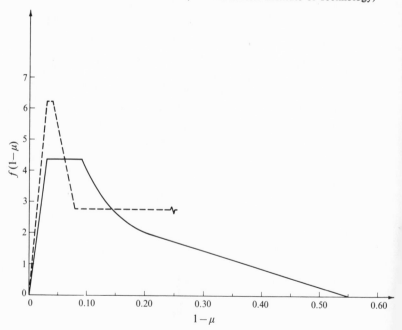

4.5×10^{17} to 2.7×10^{18} particles per second; the total number of particles emitted in the outburst shown in Figure 4.22 (at $t = -0.5 \times 10^6$ sec) is 9×10^{23}. The mass emission rate for dust $\dot{m}_d$ is $\sim 10^8$ g sec^{-1}. The gas emission rate can be calculated from the amount of gas needed to produce the observed values of v_i. The value $\dot{m}_g \approx 6 \times 10^7$ g sec^{-1} is found for times near perihelion. For a molecular weight of 25, this gives an emission rate $\dot{N}_g$ of approximately 1.5×10^{30} molecules sec^{-1}. This quantity can be estimated in a variety of ways and is an important datum in the discussion of comet models (Chapter 5).

Finson and Probstein made a partial application of their theory to Comet Van Gent (1941d) and were able to obtain rough agreement with the observed tail orientations if the dust emission occurred almost entirely before perihelion. Recently Sekanina and Miller (1973) applied the theory to Comet Bennett (1970 II) with complete success, as shown

Figure 4.24. The dust-particle-size distribution function $g(\rho_d d)$ for $Q_{PR} = 1$. (Courtesy of M. L. Finson and R. Probstein, Massachusetts Institute of Technology)

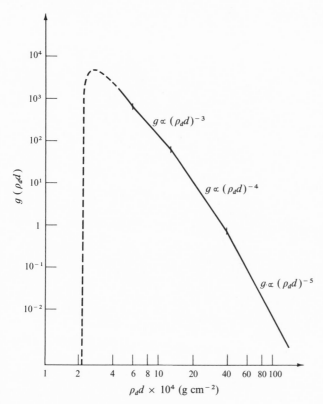

in Figure 4.25. The theory of type II or dust tails is a most valuable source of data on many aspects of cometary physics and must be regarded as one of our most secure forms of cometary knowledge.

Some comets show type II tails at large distances from the sun. The classic examples are comets Baade (1955 VI) and Haro–Chavira (1956 I), both of which had well-observed type II tails observed at heliocentric distances between 4 and 5 AU. Sekanina (1975) has shown that the

Figure 4.25. Calculated isophotes (solid curves) and measured isophotes (dotted curves) for Comet Bennett on March 18.4, 1970. See Figure 4.20 for an explanation of M and N. (Courtesy of Z. Sekanina and F. D. Miller)

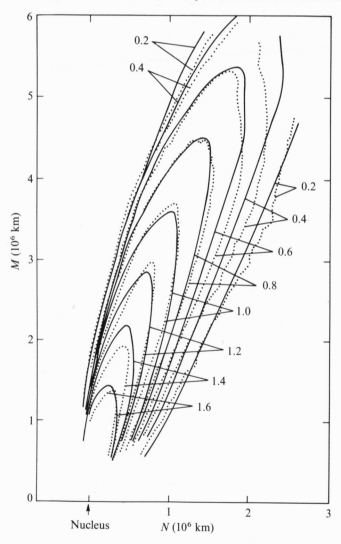

Finson–Probstein approach can explain the orientations of these tails if they are composed of particles 0.01 cm in diameter and larger. Dust particles are not known to have a size distribution with a cutoff near 0.01 cm, but such a distribution is compatible with the properties of clathrate grains as studied in the laboratory. Thus, Sekanina has suggested that the type II tails of distant comets are icy tails composed of clathrate hydrate grains. This picture requires a substance more volatile than water snow to be a major constituent of the nucleus in order to strip the grains from the nucleus at large heliocentric distances. This is quite possible for a comet that has never previously had a close approach to the sun.

Hydrogen–hydroxyl cloud

The huge hydrogen clouds now known to surround comets were predicted in 1968 by Biermann. He assumed a total production rate of 10^{30} to 10^{31} molecules sec^{-1} and, based on the ionization time scale for neutral hydrogen, estimated the cloud radius at approximately 10^7 km. In the region where both destruction of hydrogen by ionization and production of hydrogen by dissociation of parent molecules can be neglected, the density can be approximated by

$$n_H \approx \frac{Q_H}{4\pi r^2 v_H} \text{ (atoms cm}^{-3}) \tag{4.10}$$

where Q_H is the hydrogen production rate in atoms per second, r is the cometocentric distance, and v_H is the average outflow speed of the hydrogen. The outflow speed was assumed to correspond to the speed of mean thermal energy for a gas with $T = 2000$ K, a value chosen by analogy with observations of the earth's daylight-side hydrogen atmosphere. This yields $v_H \approx 6$–7 km sec^{-1}. If we also take $Q_H = 4 \times 10^{30}$ atoms sec^{-1}, equation 4.10 becomes

$$n_H \approx 5 \times 10^3 \left(\frac{10^{10} \text{ cm}}{r}\right)^2 \text{ (cm}^{-3}) \tag{4.11}$$

In the region of validity, the column density along a ray that passes the nucleus at a distance r varies as

$$N_H \approx \frac{16 \times 10^{23}}{r(\text{cm})} \text{ (atoms cm}^{-2}) \tag{4.12}$$

The value for the absorption cross section of the cometary hydrogen gas should be $\approx 10^{-13}$ cm^2. Hence, the gas has an optical depth unity where $N_H \approx 10^{13}$ atoms cm^{-2}, or at $r \approx 2 \times 10^6$ km. Biermann (1968) also found a large number of hydrogen atoms outside of this quoted distance. Thus, he concluded that "comets should, therefore, be very conspicuous objects in Ly α." The total extent of the hydrogen cloud follows from the lifetime of hydrogen for ionization τ_H (due to photoionization and charge transfer) of 10^6 sec (at 1 AU) and the speed v_H of ~ 10 km sec^{-1} and is $\sim 10^7$ km. The radiation pressure due to the solar Lyman α was found to be large enough to expect considerable asymmetry in the outer shape of the hydrogen cloud. Thus, Q_H, v_H, τ_H, and the effects of radiation pressure need to be specified in order to calculate a model of cometary hydrogen distribution.

On January 14, 1970, the first ultraviolet observations of a comet were made from the second Orbiting Astronomical Observatory (OAO-2) when the spectrum of Comet Tago–Sato–Kosaka (1969g) was recorded. The observations clearly showed the huge cloud of Lyman α emission. Observations of the comet were continued throughout January 1970. Comet Bennett (1969i) was observable from space in February 1970, and it was observed from OAO-2 (Figure 4.26) and by two experiments on board the fifth Orbiting Geophysical Observatory (OGO-5). The OGO experiments were wide-field photometers that were more sensitive than the OAO-2 instrument. In addition, OGO-5 was at a high altitude, above most of the earth's hydrogen cloud (geocorona). As a result, it was possible to observe much fainter emission with OGO-5 than with OAO-2. Comet Encke was detected by OGO-5 in January 1971, but Comet Toba was not. Extensive observations of the hydrogen cloud surrounding Comet Kohoutek were obtained in 1973–74 (see Chapter 7).

A spectral scan of Comet Bennett is shown in Figure 4.25, with hydroxyl (OH) and oxygen (O I) emission in addition to Lyman α. Comparison of the different intensities leads to the conclusion that H_2O is the likely ultimate parent molecule for H, OH, and O; the O and some of the H would result from the dissociation of OH.

The terrestrial neutral hydrogen above the OAO-2 spacecraft had an optical depth in Lyman α of approximately 0.6 and a width (due to the thermal Doppler effect) corresponding to a temperature of approximately 1000 K. The relative velocity between the earth and Comet Tago–Sato–Kosaka varied in such a way that the cometary Lyman α emission line moved in wavelength across the telluric absorption as the comet passed perigee. Analysis of the decrease in observed Lyman α emission gives a Doppler width corresponding to a temperature of 1600 K for the cometary hydrogen.

Clearly, the central part of the cometary hydrogen clouds is optically thick, and detailed radiative transfer treatments are necessary for the interpretation of the multiple scattered radiation. Monte–Carlo investigations have been carried out and the results applied to Comet Bennett.

Figure 4.26. Observations of Comet Bennett from the Orbiting Astronomical Observatory (OAO-2). (*a*) Isophotes in Lyman α on April 16, 1970. The arrow indicates the sunward direction. (*b*) Spectral scans. (Courtesy of A. D. Code, T. E. Houk, and C. F. Lillie)

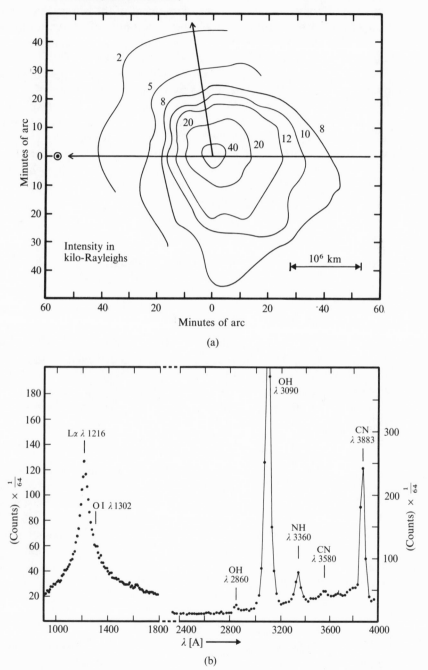

Also, note that the region of maximum Lyman α intensity may not coincide with the cometary nucleus. Treatment of the outer part of the hydrogen cloud is relatively simple because only single scattering is involved. Much of our quantitative data on the hydrogen cloud refers to the outer regions and follows the general outline given below.

The model of the hydrogen cloud we shall describe is the fountain model (see Chapter 2) as generalized by Haser to include the effects of solar radiation pressure, lifetimes of molecules, and different velocity distributions for the outflowing molecules. We stress again that this model strictly applies only to the optically thin outer region, typically beyond cometocentric distances of 10^6 km. To proceed, we assume an isotropic point source that produces neutral hydrogen atoms at the rate Q_H. These atoms leave the source with speeds corresponding to a Maxwellian distribution, namely,

$$f(v) \, dv = \frac{4v^2}{\pi^{1/2} \, v_0^3} \exp\left[-(v/v_0)^2\right] dv \qquad (4.13)$$

Here v_o is the most probable speed that is related to the mean speed by

$$\langle v \rangle = \frac{2}{\pi} \, v_o \qquad (4.14)$$

The quantity $\langle v \rangle$ is usually identified with the outflow speed of the hydrogen atoms v_H. Neutral hydrogen atoms are destroyed (ionized) by photoionization or charge exchange with solar wind protons. At 1 AU the photoionization lifetime is $t_{rad} = 1.4 \times 10^7$ sec and the solar wind charge-exchange lifetime is $t_{sw} = 3 \times 10^6$ sec. The total lifetime of a hydrogen atom, t_H, is given by

$$t_H = (t_{rad}^{-1} + t_{sw}^{-1})^{-1} \qquad (4.15)$$

or $t_H = 2.5 \times 10^6$ sec. This value is dominated by the solar wind charge-exchange lifetime. At distances other than 1 AU the lifetime can be calculated using the r^{-2} law. In practice t_H is an empirically determined parameter.

Additional assumptions used in the calculations are: Q_H is constant, phase effects are negligible, and resonance scattering in Lyman α is isotropic. Then, the emission from the optically thin parts of the hydrogen cloud is given by

$$4\pi I_c = \left(\frac{\pi e^2}{m_e c}\right) f N F_\odot \qquad (4.16)$$

The intensity $4\pi I_c$ is usually quoted in rayleighs [10^6 photons (cm^2-sec)$^{-1}$], the oscillator strength $f = 0.42$, $\pi e^2/m_e c$ are the usual atomic parameters, N is the column density, and $F_\odot$ is the solar flux at 1 AU estimated at 3.2×10^{11} photons (cm^2 sec-A)$^{-1}$. This flux gives a radiation pressure due to Lyman α of $b = 0.57$ cm sec^{-2}, which is nearly equal to the solar gravity of 0.6 cm sec^{-2}.

Haser's density integral can be evaluated as follows. We introduce a coordinate system with the origin in the nucleus, the sun in the direction of the negative z axis, and the earth in the direction of the positive x axis. The space density near the nucleus is given by an equation analogous to equation 4.10, that is,

$$n_H(x, y, z) = \frac{Q_H e^{-t/t_H}}{4\pi v_H (x^2 + y^2 + z^2)} \tag{4.17}$$

where t is the time of travel. Away from the nucleus, the atoms are accelerated in the positive z direction. The effect is a density distribution with symmetry around the z axis rather than the spherical symmetry of equations 4.17 and 4.10. The trajectories of the hydrogen atoms are parabolas and, for each energy, a point in the hydrogen atmosphere can be reached via two separate trajectories. Thus, two components, denoted by $\pm$ in the following equation, must be added to obtain the total density. The result is

$$n_H(x, y, z) = \frac{Q_H}{8\pi v_H} \{[a \pm (x_o^2 - x^2)^{1/2}](x_o^2 - x^2)^{1/2}\}^{-1}$$
$$\times \exp\left\{-\frac{1}{t_H}\left(\frac{2}{b}\right)^{1/2}[a \pm (x_o^2 - x^2)^{1/2}]^{1/2}\right\} \tag{4.18}$$

where

$$a = z + \frac{v^2}{b} \tag{4.19}$$

and

$$x_o = [a^2 - (z^2 + y^2)]^{1/2} \tag{4.20}$$

Recall that b is the acceleration due to radiation pressure and that v is the initial velocity of the hydrogen atoms. The upper bound to the integration is x_o, which is determined by the maximum value of v considered (see equations 4.19 and 4.20); note that x_o is a function of y and z and forms a rotational parabaloid about the x axis. The column density is then determined by twice the integral from $x = 0$ to $x = x_o$ or

$$N(y, z) = 2 \int_0^\infty f(v) \int_0^{x_o} n_H(x, y, z) \, dx \, dv \tag{4.21}$$

Here $f(v)$ is obtained from equation 4.13 or its equivalent under other assumptions, and $n_H(x, y, z)$ is composed of the two parts of equation 4.18.

Integrations such as these were carried out by Bertaux, Blamont, and Festou (1973) and by Keller (1973). The solutions are determined by empirically varying Q_H, v_H, and t_H until a fit to the isophotes is obtained. Sample results are shown in Figure 4.27; in addition, we have recent results for Comet Kohoutek (see Chapter 7). An average hydrogen atom traveling sunward with initial kinetic energy per unit mass $(v_H^2/2)$ against the solar radiation pressure b has a speed that decreases to zero when the work done $z_{max} b$ equals the initial kinetic energy. Therefore, we have

$$z_{max} = v_H^2/2b \tag{4.22}$$

Thus, the sunward extent of the hydrogen cloud and the isophotes on the sunward side should be determined primarily by v_H, and empirically this is found to be true. Hence, v_H is first determined from the sunward or forward intensity profile, and then t_H is determined from the antisolar or backward intensity profile. The intensity profile perpendicular to the radius vector can be used as a check on consistency. The fits to the observed profiles are not perfect, but they are certainly adequate to determine values of Q_H, v_H, and t_H with reasonable accuracy. The observed intensity profiles require a distribution of initial injection speeds (such as equation 4.13) and are not compatible with monokinetic injection.

The t_H values determined are close to the values expected on theoretical grounds (quoted above). The empirically determined values of Q_H are consistent with values determined by other independent approaches. H. U. Keller (1973) has applied this technique to Comet Bennett (1969i) and has found the results shown in Table 4.3. These

Table 4.3. *Hydrogen production parameters in Comet Bennett*

r(AU)	v_H (km sec^{-1})	Q_H(atom sec^{-1})	t_H(sec)
0.61	7.9	1.2×10^{30}	2.5×10^6
0.62	7.9	1.0	2.5
0.70	7.9	0.92	2.0
0.73	7.9	0.79	2.0
0.78	9.0	0.77	2.5
0.80	7.9	0.69	2.0
0.82	9.0	0.79	2–2.5
0.86	7.9	0.80	2.0

Source: Keller (1973).

Figure 4.27. Comparison of observed and computed Lyman α isophotes for Comet Bennett, April 1, 1970. (*a*) Observed by J.-L. Bertaux and J. Blamont. (*b*) Calculated. (kR = kilo-Rayleigh). (Courtesy of H. U. Keller, Max-Planck-Institut für Aeronomie)

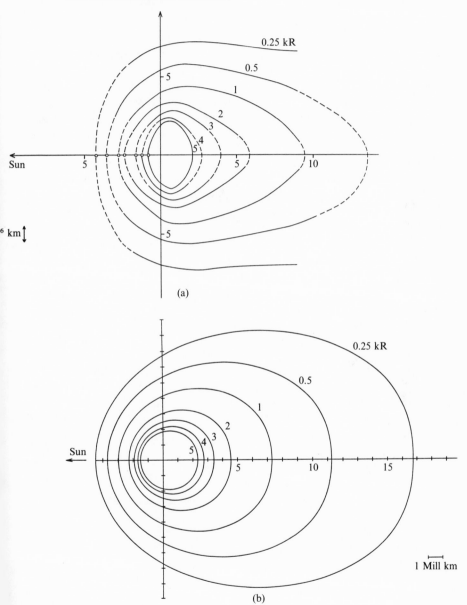

results are consistent with a variation of Q_H with heliocentric distance as $r^{-1.5}$, with an error in the exponent of about ± 0.5. Other comets show variations in the range $r^{-1.5}$ to $r^{-2.0}$. Compare these results with those obtained for other comets as given in Table 4.4.

Additional data concerning the cometary environment can be obtained from the Lyman α intensity contours.

1. The curvature of the extended atomic-hydrogen tail gives a value for the radiation pressure using the standard syndyne formulas. This method uses only the positions of intensity maxima, and hence it is not subject to the usual uncertainties associated with absolute calibrations. The values found are compatible with the values quoted here, but are somewhat higher.

2. Because the lifetime t_H is determined primarily by the proton flux in the solar wind, its determination provides an opportunity to monitor the proton flux. A possible variation was found: The solar wind flux may be higher at solar latitudes poleward of $\pm 45°$.

The theory of the coma (described in a later subsection) and observational checks indicate that the overall outflow speed from a clathrate hydrate nucleus should be ≈ 0.5 km sec^{-1} or less than 1 km sec^{-1}. This speed refers to the region relatively near the nucleus where collisions between atoms and molecules are important, that is, a thermalization region. If some dissociation occurs in the collision region, the outflow speed could be increased to (say) 2 or 3 km sec^{-1}, but probably no more. Thus, the observed v_H of 8 km sec^{-1} is probably related not to details of the flow but to the energy budget of the photodissociation reaction that produces the hydrogen, outside the thermalization region. If H_2O were the primary parent molecule, a speed of 19 km sec^{-1} would be expected, and clearly this value was too high. Other candidates for the key parent molecule were OH or other hydrocarbons. Because OH

Table 4.4. *Hydrogen production rates in different comets at (or reduced to) 1 AU*

Comet	Q_H (atom sec^{-1})
Bennett (1970 II)	4×10^{29a}
Kohoutek (1973 XII)	$2.5-4 \times 10^{28}$
Tago–Sato–Kosaka (1969 IX)	8×10^{29}
P/Encke	3×10^{27}
West (1975n)	5×10^{29}

[a] These values are approximate.

can be produced by the photodissociation of H_2O, it became the likely candidate. In this picture, the hydrogen produced by the photodissociation of H_2O are thermalized near the nucleus. The additional hydrogen produced by the subsequent photodissociation of OH would originate outside the thermalization region and hence could reach the outer parts of the hydrogen cloud where the 8 km sec^{-1} value for v_H is determined. Observations of the OH intensity in Comet Kohoutek have apparently confirmed the validity of this general model.

Heliocentric brightness variations

In Chapter 2 we developed the theory for the heliocentric variation of cometary brightnesses based on the adsorption–desorption model of B. U. Levin. An interesting conclusion concerning the asymptotic behavior of this variation comes from the picture of cometary phenomena, as outlined in the previous section.

The total brightness of comets should vary as r^{-2} for a heliocentric distance small enough that essentially all the solar radiant energy incident on the comet nucleus goes into gas production. Following a simple line of argument developed by M. Mumma (private communication), the intensity of light radiated by a kind of atom, molecule, or dust particle is

$$I \sim \sigma F N \qquad (4.23)$$

where I is the intensity emitted in all directions, σ is the cross section for scattering, F is the solar flux in the frequency band of interest, and N is the number of scatterers. (In a situation where more than one species contributes to the intensity, equation 4.23 will be a sum of similar terms. The generalization of the remainder of the argument is quite simple.) The number of atoms, molecules, or dust particles is clearly

$$N \sim Q\tau \qquad (4.24)$$

where Q is the production rate and τ is the mean or effective (for dust) lifetime. Thus,

$$\begin{aligned} I &\sim (\sigma F)\,(Q\tau) \sim (\sigma F\tau)_o r^{-2} Q r^{+2} \\ &\sim (\sigma F\tau)_o \times Q \end{aligned} \qquad (4.25)$$

This reduction assumes (quite reasonably) that the lifetime of species is controlled by solar radiation, which falls off as r^{-2} (this includes radiation pressure for dust particles). Therefore, if the total incident solar radiant energy goes into gas production, we find

$$I \sim (\sigma F \tau Q)_o r^{-2} \qquad (4.26)$$

or

$$I \propto r^{-2} \tag{4.27}$$

In addition, the total cometary brightness at large distances from the sun when there is little or no gas production obviously varies as r^{-2}. Hence, the total brightnesses of comets should vary as r^{-2} at both large and small heliocentric distances and should vary as a higher inverse power at intermediate distances where gas production has begun but does not utilize all the radiant energy. Observations of Comet Encke (Figure 4.28) over several apparitions clearly show this variation.

Figure 4.28. The brightness of Comet Encke versus heliocentric distance. The solid curves labeled $n = 2$ correspond to the r^{-2} variation described in the text. (Courtesy of M. Mumma, NASA–Goddard Space Flight Center)

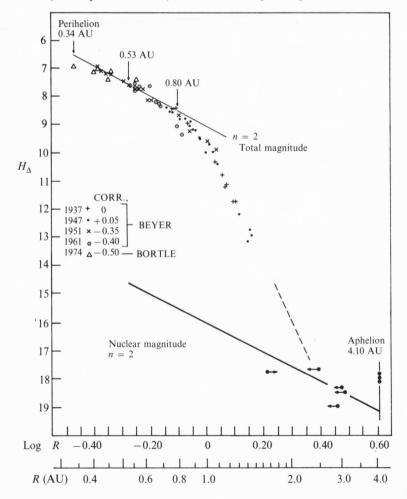

Coma

The coma is composed of neutral molecules and dust particles in an essentially spherical volume centered on the nucleus. Molecules in the gaseous coma that have been identified spectroscopically are listed in Table 4.1. These molecules probably have their origin in the clathrate ices of the nucleus.

Coma sizes range up to 10^5 or 10^6 km and typically have a maximum when the comet is located between 1.5 and 2.0 AU. For most comets, as they approach nearer the sun than this distance, their comas are observed to contract. The comas are apparently quite small at large distances from the sun, that is, at distances greater than 3 or 4 AU. As a matter of terminology, the coma does not include tail molecules passing through.

The coma gases flow away from the nucleus at approximately 0.5 km sec^{-1}. This rate can be established from the motion of expanding rings or halos, from the speed required to explain the Greenstein effect, and from the theory of coma flows, including the kinetic theory of those gases vaporizing at nuclear temperatures and the speed required to drag dust particles up to the speed required to explain the widths of type II (dust) tails.

A great deal of the work on cometary comas consists of the observation and interpretation of intensity contours of radiation from one given constituent. Narrow passband filters are available that isolate the molecular bands of one molecule, for example, C_2 or CN. The situation has some similarities with the interpretation of intensities from the hydrogen cloud (discussed in the previous subsection). The molecules under discussion are assumed to originate from a spherical volume surrounding the nucleus with a speed determined by flow conditions near the nucleus, and they are assumed to photodissociate because of the solar radiation field. The mass of the comet has no significant decelerating effect on the molecules (outside of the near-nuclear volume), and hence the flow is at constant speed, v_0. The time scale for photodissociation is defined such that the change in density caused by photodissociation is given by

$$\frac{dN}{dt} = -\frac{N}{\tau_0} \tag{4.28}$$

where N is the number density and τ_0 is the time for the number density to fall to $1/e$ of its initial value. The distance traveled by the average particle before dissociation is given by

$$R_0 = \tau_0 v_0 \tag{4.29}$$

Dissociation alone would produce a density varying as e^{-r/R_0}, and the equation of continuity for a spherical system and constant speed produces an r^{-2} variation. Hence, this simple picture produces a total variation as[1]

$$N(r) = \left(\frac{R}{r}\right)^2 N(R)e^{-r/R_0} \qquad (4.30)$$

where R is the radius of the nucleus. This equation is essentially the same as equation 4.17, but the situation in the coma is probably more complicated. The simple molecules that we observe are thought to originate from photodissociation of relatively complex parent molecules. Equation 4.30 can be generalized in a straightforward manner to include two decay processes, and we have

$$N(r) = \left(\frac{R}{r}\right)^2 N(R)[e^{-r/R_0} - e^{-r/R_1}] \qquad (4.31)$$

Here R_0 refers to the daughter molecules and R_1 to the parent molecules. In the literature these processes are often discussed in terms of $\beta_0 = (R_0)^{-1}$, and so on. Equation 4.31 reduces to equation 4.30 for $R_1 = 0$ or for $\beta_1/\beta_0 = \infty$.

Equation 4.31 must be integrated to obtain the column density along the line of sight that passes the nucleus at the closest distant ρ, and Haser has obtained

$$N(\rho) = 2N(R_m)R_m^2 \frac{\beta_0\beta_1}{\beta_1 - \beta_0} e^{\beta_0 R_m} \frac{1}{\beta_0\rho}[B(\beta_0\rho) - B(\beta_1\rho)] \qquad (4.32)$$

Here R_m is the distance at which the expression in brackets in equation 4.31 is a maximum, $N(R_m)$ is the density at that point, and

$$B(z) = \frac{\pi}{2} - \int_0^z K_0(y)\,dy \qquad (4.33)$$

where $K_0(y)$ is the modified Bessel function of zero order and the second kind. Thus, the surface brightness $S(x)$ can be expressed in a dimensionless form[2] as

$$S(x) \propto N(\rho) \propto \frac{1}{x}\left[B(x) - B\left(\frac{\beta_1}{\beta_0}x\right)\right] \qquad (4.34)$$

where $\beta_0\rho = x$.

Physically, the preceding discussion can be summarized in terms of the schematic diagram shown in Figure 4.29, which is a plot of $\log S$ versus $\log \rho$. If there is only outflow at constant velocity (no creation or destruction processes), the slope of the brightness curve in this diagram would be -1. If the brightness curve falls slower than -1, the molecules are being created (near the nucleus). If the brightness curve falls faster than -1, the molecules are being destroyed (far from the nucleus).

A comparison of the model calculations versus observations of C_2 emission are shown in Figure 4.30. A reasonably good fit is shown for $\beta_1/\beta_0 = 9$, whereas $\beta_1/\beta_0 = \infty$ does not fit. Equation 4.31 is correct only if the daughter molecules move radially and the parent molecules all decay at about the same distance from the nucleus. If these conditions are not met, equation 4.31 must be replaced by a double integral. In his Ph.D. thesis, Festou (1978) made the double integration in one limited case and found significant departures from Haser's model. More detailed mathematical treatment of the problem is in order. Recently, Combi and Delsemme (1980) replaced Haser's simplified model with an average random-walk model. They showed that measured radial scale lengths are lower limits to a range of possible true nonradial scale lengths.

Density determinations of the type just discussed apply to the outer, optically thin parts of the coma. However, in 1965 (at the thirteenth International Astrophysical Colloquium, Liège) evidence supporting important, physically distinct regions within the coma began to accumulate. Although it had been recognized previously that an optically thick region existed in the coma, few workers envisioned a region in the coma with densities high enough to produce collision effects in molecu-

Figure 4.29. Schematic variation of column density of a species with both creation and destruction mechanisms. See text for discussion.

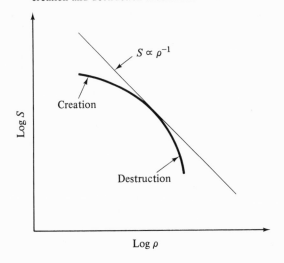

lar spectra or chemical reactions. Thus, the view began to change in 1965 based on suggestions by Delsemme (1966), Jackson and Donn (1966), and Malaise (1970).

The coma could be divided into three physically distinct regions along the lines suggested by Malaise (1970).

1. A nuclear region is formed by an optically thick dust cloud. The radius of this region is given by the distance from the nucleus at which the optical depth is unity in the continuum, typically a few hundred kilometers. This region may be responsible for the false or photometric nuclei of comets, and it also produces the reflected solar spectrum.

2. The cometary atmosphere is the region outside the nuclear region where collisions are important. The outer boundary of this region is the point where the tangential mean free path is

Figure 4.30. Comparison of theory and observation (solid curve) for the surface brightness variation in the C_2 bands. The dot-dashed curve (which is coincident with the dashed curve at upper left) is the simple $S \propto \rho^{-1}$ variation. The dashed curve ($\beta_1/\beta_0 = \infty$) is a fit with destruction processes only. The dotted curve ($\beta_1/\beta_0 = 9$) includes both creation and destruction processes; it is a good fit. (After C. R. O'Dell and D. E. Osterbrock)

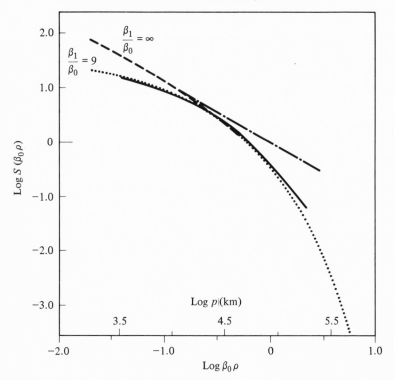

equal to the scale height or radial distance over which the gas density falls by a factor e. This level in planetary atmospheres is the so-called critical level or exospheric base.

3. The coma proper is the region beyond the atmosphere where collisions are unimportant; this is equivalent to the exosphere of planetary atmospheres. In comets, molecules in the coma are excited only by radiative processes.

Convincing evidence for region two, the atmosphere, was presented by Malaise (1970). He analyzed high-dispersion spectra of several comets and computed detailed synthetic spectra. Good agreement could not be obtained with purely radiative excitation; collisional excitation was required. Besides the atomic parameters of the molecule (here CN) under consideration, the calculation of synthetic spectra requires at least the following quantities: (1) detailed knowledge of the solar spectrum as seen by the comet; (2) specification of the comet's radial velocity with respect to the sun (Swings effect); (3) the expansion velocity of the cometary gas v_e; differential motions in the cometary gas itself due to expansion produces spectral effects (Greenstein effect); (4) a parameter giving the relative efficiency of collisions and fluorescence in determining the populations of levels in the molecule. Define the parameter α by

$$\frac{\alpha}{1 - \alpha} = \frac{\tau_{fl}}{\tau_{coll}} = Q \tag{4.35}$$

where τ_{fl} and τ_{coll} are the characteristic times for establishing a fluorescence or collisional (Boltzmann) equilibrium, respectively, from an arbitrary initial distribution. An α of 1 (or 100%) implies a collisionally dominated regime, and an α of 0 implies a regime dominated by radiation. The collision time is taken as

$$\tau_{coll}^{-1} = 2N\sigma \frac{2\pi kT}{\mu} \tag{4.36}$$

where N is the total density in the model atmosphere, σ is the cross section for collisionally exciting the rotational transitions, and $\mu[=m_1 m_2/(m_1 + m_2)]$ is the reduced mass of the colliding particle. The fluorescence time is determined by absorption coefficients and the selection rules for radiative transitions. At 1 AU from the sun, $\tau_{fl} \sim 10^2$ sec. We have

$$Q = 2.3 \times 10^{12} N\sigma\tau_{fl} \frac{T^{1/2}}{A} \tag{4.37}$$

where A is the reduced mass in atomic units, T is the temperature in K, and the collisional excitation cross section is in units of 10^{-6} cm^2.

(5) Finally, a parameter describing the photodissociation of the molecule (CN) being studied must be specified; the distance traveled by a molecule before dissociation is called D.

Malaise (1970) has carried out the quite complex calculations of synthetic spectra for CN with the effects of the instrumental profile taken into account. The spectra used were taken at dispersions of 19.5 Å mm^{-1} and 39.0 Å mm^{-1}. The main parameters that determine the shape of the spectra are summarized in Table 4.5. The subscript 0 refers to the value of the specific parameter at 10^4 km from the nucleus; T, v_e, and D are constant throughout the atmosphere. The Greenstein effect is shown by the values in the column marked v_e. The Swings effect can be illustrated by the spectra shown in Figure 4.3. Clearly, models with purely radiative excitation cannot always reproduce the observed spectra. Some collisional effects are definitely present. At 10^4 km from the nucleus the collision frequency is $\sim 10^{-3}$–10^{-2} sec^{-1}. Although collisional excitation can be important in the central regions of cometary atmospheres, it is important to remember that overall, the influence of collisions is relatively small and that fluorescence is still the dominant excitation process.

When the nature of the exciting particles is specified, their density can be calculated from the known values of $N_0\sigma A^{1/2}$. Electrons and protons are readily excluded as the exciting particles; the evidence indicates that the exciting particles are the total ensemble of radicals and molecules in the cometary atmosphere. The total density is four to five orders of magnitude higher than the densities calculated from observations of visible species, such as CN. (In fact, they are probably too high, due to uncertainties in cross sections.) The total atmospheric densities are in the range 10^7–10^8 cm^{-3} at 10^4 km from the nucleus for a comet at 1 AU or less. Because we have estimates of both total density

Table 4.5. *Parameters obtained from the synthetic spectra of comets*

Comet	r (AU)	α_0 (%)	T (K)	$N_0\sigma A^{1/2}$ $\times 10^{-8}$	v_e (km sec^{-1})	D (10^5 km)
Seki–Lines	0.55	90	600	48	0.80	4.7
Ikeya	0.66	90	580	33	0.74	6.0
Encke	0.69	55	480	5.7	—	—
Honda	0.71	20	550	1.0	—	—
Ikeya	0.74	(75)	500	(10)	0.67	6.4
Encke	0.79	25	350	1.5	—	—
Burnham	0.99	<2	(350)	<0.1	0.59	2.2
Candy	1.15	20	325	0.5	—	—

Source: After Malaise (1970).

and outflow speed, the total efflux is roughly known and amounts to between 10^{30} and 2×10^{31} molecules sec^{-1}. It is reassuring to note that these total effluxes are close to the values calculated from studies of the hydrogen cloud and from studies of the dust tails of comets. Arpigny (1965) has also carried out synthetic spectrum calculations for a number of molecules. He has reviewed the calculations of Malaise and has suggested that they should be redone using a better statistical equilibrium equation and improved molecular parameters, more accurate wavelengths, and including additional lines in the CN spectrum. Arpigny concludes that Malaise's calculations were as good as the state-of-the-art when they were made. However, the importance of CN is such that improved calculations are warranted.

We now turn to a discussion of the expansion velocities found in cometary comas. According to our definition of the cometary atmosphere, the kinetic mean free path is small and the problem can be treated hydrodynamically. Probstein (1968) has considered the problem of the expansion of a two-phase "dusty gas." In essence, the gas sublimates from the nucleus, expands outward, and drags the liberated dust particles along with it. The dust–gas coupling was computed using standard free-molecular drag coefficients, and the details of the expansion depend on the amount of coupling between the dust and the gas. Generally, the solution contains a sonic or critical point at which the subsonic solution (valid near the nucleus) crosses over to a supersonic solution (valid away from the nucleus). This is the type of transonic flow pattern found in the solar wind and in rocket engines (the de Laval nozzle). We do not discuss the basic physics of such flows here. However, the gas exhaust speed from a de Laval nozzle is $3^{1/2}v_s$, where the sound speed $v_s = (\gamma P/\rho)^{1/2}$. For $T = $ K, $\gamma = 1.4$, and mean molecular weight of 20, $3^{1/2}v_s \approx 0.6$ km sec^{-1}. The presence of dust in the gas reduces the terminal speed (see Figure 4.31). The numerical values quoted here should not be taken too seriously, but clearly these flows can approximately reproduce the injection speeds required by type II tails.

The terminal speed of the dust particles is reached within approximately 20 radii of the nucleus or within some 20 to 100 km. Thus, the terminal speed is reached well within the nuclear region of the coma and, so far as calculations of the structure of dust tails are concerned, the dust can be considered as emanating from a point source.

The terminal speed v_i was found by Probstein to be expressible in the form

$$\frac{v_i}{(c_p T_o)^{1/2}} = g(M, \beta) \tag{4.38}$$

where T_o is the temperature of the gas at the nucleus and c_p is the specific heat of this gas. The parameters M and β are

$$M = \dot{m}_d / \dot{m}_g \tag{4.39}$$

and

$$\beta = \frac{16}{3}\, \pi \rho_d\, dr_o (c_p T_o)^{1/2} / \dot{m}_g \tag{4.40}$$

In these equations, $\dot{m}_g$ and $\dot{m}_d$ are the mass flow rates of the dust and gas, respectively; ρ_d is the density of the dust; r_o is the radius of the nucleus; and d is the diameter of the dust particles. The function $g(M, \beta)$ is given in Figure 4.31. The parameter β is seen to represent a drag. For an assumed T_o of 200 K, we have $(c_p T_o)^{1/2} = 0.48$ km sec^{-1}. Finson and Probstein (1968) found a range of v_i between 0.2 and 0.4 km sec^{-1} by fitting the values required to explain the widths of Comet Arend–Roland's dust tail. The value $v_i = 0.33$ km sec^{-1} was determined for most of the orbit near perihelion, and the value $v_i/(c_p T_o)^{1/2} = 0.33/0.48 = 0.69$ is plotted in Figure 4.31. A value of β in the approximate range 0.6 to 25 is indicated. Equation 4.40 can be utilized to calculate a value for $\beta \dot{m}_g$; for $r_o = 5$ km, $T_o = 200$ K,

Figure 4.31. Terminal speeds of dusty gas flows $v_i/(c_p T_o)^{1/2}$ versus β. The dashed lines at the left-hand edge are values for $\beta \to 0$. Note the decrease in speed as $M = \dot{m}_d/\dot{m}_g$ increases. (Courtesy of R. F. Probstein, Massachusetts Institute of Technology)

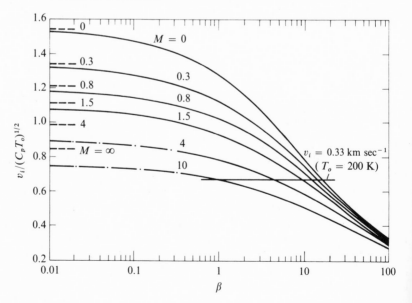

$d = 1$ μm, and $\rho_d = 3$ g cm^{-3}, we find $\beta \dot{m}_g \approx 1 \times 10^8$ g sec^{-1}. The values of β inferred from the v_i required for Comet Arend–Roland then give gas mass loss rates $\dot{m}_g$ of 0.8×10^7 g sec^{-1} to 3×10^8 g sec^{-1}. For a mean molecular mass of 20, these numbers imply losses in the range 3×10^{29} molecules sec^{-1} to 1×10^{31} molecules sec^{-1}. The order of magnitude of these results is entirely compatible with the results obtained by other methods. The ratio $\dot{m}_d/\dot{m}_g$ is not well determined (see Figure 4.31). However, observational estimates for the dust production rates by Liller (1960) for comets Arend–Roland and Mrkos are approximately 10^8 g sec^{-1}. Hence, the evidence is consistent with $M \sim 1$ and gas and dust outflow rates $\sim 10^8$ g sec^{-1}.

The results discussed above require modification for the case of large dust particles. The usual (micron size) particles that form the ordinary type II tail are not significantly influenced by the gravitational attraction of the nucleus, but the larger particles are. The size of the dust particles enters parameter β through equation 4.40. For small values of β, the ratio of the dust speed to the thermal speed in the gas v_d/v_g is nearly constant or independent of β. For large values of β, v_d/v_g is approximately $(9/\beta)^{1/2}$. When the approximation for large values of β are modified to include the gravitational attraction, we have

$$v_d{}^2 = \frac{9v_g{}^2}{\beta} - \frac{2GM_c}{r_o} \tag{4.41}$$

where M_c is the mass of the comet. If the gas mass loss rate $\dot{m}_g$ is known, we can estimate $(\beta/d\rho_d)$ from equation 4.40. Then, the size of the largest particles that can escape is given by putting $v_d = 0$ in equation 4.41. Gary and O'Dell (1974) find from studies of Comet Kohoutek that the maximum particle size to escape is roughly 10^2 larger than the size that dominates type II tails. We return to this subject in Chapter 7.

Cometary outbursts

Dusty flows may also become anomalously large for some reason, leading to an outburst in a comet. Sooner or later, our understanding of comets must include a detailed physical explanation of these outbursts, which are usually sudden increases in the total cometary magnitude by two to five magnitudes, corresponding to an increase in luminosity of 6 to 100. Larger increases in luminosity have been recorded. The typical outburst lasts roughly 3 to 4 weeks. Approximately three dozen prominent comets have outbursts, the most famous being P/Schwassmann–Wachmann, shown in Figure 4.32. Outbursts appear to occur throughout the inner solar system, with no strong dependence on heliocentric distance or whether before or after perihelion. Because

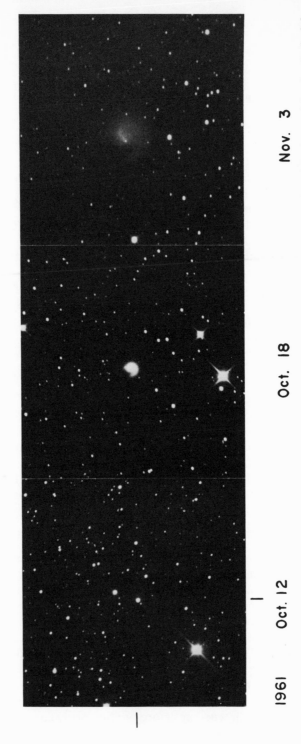

1961   Oct. 12   Oct. 18   Nov. 3

Figure 4.32. An outburst of Comet Schwassmann–Wachmann I. The reproductions have been matched in scale, orientation, and density. In the left panel, the starlike comet image is indicated by the lines. The comet's orbit is nearly circular at a heliocentric distance of 6 AU. (Courtesy of E. Roemer, University of Arizona; official U.S. Navy photograph)

of the increase in brightness, a number of comets have been discovered during an outburst.

A manifestation of the outburst process is a shell of material (not always symmetrical) that expands at speeds in the range 100 to 500 meters sec^{-1}. This planetlike disk can reach 2' to 3' in diameter in about a month; then it fades. Note that the increase in area of the disk does not account for the increase in luminosity. The observational evidence is consistent with the light originating from sunlight scattering off dust particles (diameters ~ 0.5 μm).

Several mechanisms have been suggested for outbursts. These include, as listed by Hughes (1975) in his review, pressure release from gas pockets, explosive radicals, amorphous ice, impact cratering by boulders, breakup of the nucleus, and nuclear crushing; Figure 5.5 illustrates some of the possibilities. Note that the cause of outbursts need not be a single mechanism.

There is also evidence that the nucleus is surrounded by a halo of icy grains. We defer consideration of the icy grain halo (IGH) until we discuss the model of the nucleus suggested by Delsemme, to which it is closely related (see Chapter 5).

Nucleus

Strong indirect information points to the existence of a cometary nucleus, but direct information concerning radius and mass is very difficult to obtain. Discussion of the chemical composition and physical nature of the nuclear material is deferred until we discuss the overall comet model (Chapter 5).

The most reliable estimates of cometary radii have been made by Roemer (1963). Except under very special circumstances, it is not possible to resolve the nucleus directly. Rather, the sizes are inferred by the brightness of the nucleus with the aid of assumptions concerning the nuclear albedo and phase function. Successful application of this technique requires the separation of the unwanted light reflected by the inner coma from the light reflected by the nucleus. Roemer has stressed that this separation can be successfully achieved only by the use of long-focus telescopes with adequate scale. The 40-in astrometric reflector of the U.S. Naval Observatory's Flagstaff Station produces nearly "stellar" images of the cometary nucleus although recording little trace of the coma.

Houziaux's formula can then be used to convert the estimates of cometary magnitude m_c into cometary radii:

$$m_c = m_\odot - 5(\log R - \log r - \log \Delta) - 2.5[\log A + \log \phi(\theta)] \quad (4.42)$$

Here, $m_\odot$ ($= -26.72$) is the magnitude of the sun; R is the radius of the nucleus (in AU); r is the heliocentric distance; Δ is the geocentric distance; A is the albedo; and $\phi(\theta)$ is the phase function. The radii were calculated for albedos of 0.02 and 0.7, values that correspond to the extreme values for other objects in the solar system. If the icy-conglomerate model of the nucleus is correct, radii derived from the higher albedo might be appropriate. For nearly parabolic comets, the majority of the radii fall in the range 1 to 10 km. For periodic comets, the radii are approximately 1 km or smaller. These values are an order of magnitude smaller than other values found in the literature, and the difference serves to underscore the need for observational precautions.

Occasionally comets pass close to the earth, and the circumstances allow somewhat more direct checks on cometary radii. On June 27, 1927, P/Pons–Winnecke passed within 0.039 AU of the earth. Baldet observed the comet under extremely favorable conditions and found a disk with a diameter of 0.3″; this directly determines a *maximum* diameter of 5 km. Calculations similar to those described above for different albedos give diameters in the range 200 to 620 meters.

We suggest a simple way to estimate the masses of comets. The range of radii seems established, and if the density is assumed to be 1 g cm^{-3}, the mass is roughly $M_c \sim 4\, r_c^3$. For r_c in the range 1 to 10 km, the masses are in the range 4×10^{15} g to 4×10^{18} g.

Directly determined masses of comets are extremely difficult to obtain. In Chapter 2 we described upper limits based on the failure to find the gravitational effect of comets on other bodies in the solar system. These values are almost surely much too high. The splitting of comets may provide an opportunity to estimate masses. Stefanik (1966) has studied the splitting of 13 comets and finds an average separation speed of 20 meters sec^{-1}. If one arbitrarily assumes this speed to be the escape speed, results consistent with a radius of 20 km, a density of 1 g cm^{-3}, and a mass of 3×10^{19} g are obtained. The mass and radius quoted seem high, particularly if we consider an icy nucleus with a density close to 1 g cm^{-3}. If one looks at Stefanik's results carefully, one can conclude that they have little physical meaning. The average separation rates were measured for distances at which nuclear gravity is small or negligible, and there is no reason to suppose that the measured speeds represent escape speeds. The numerical values should be viewed accordingly.

Sekanina (1977) has recently proposed a new model for the motions of fragments of split comets. In the model he suggests that fragments of a split comet drift apart because they experience slightly different non-gravitational forces. This concept is in contrast to the earlier idea that the process of splitting itself exerted impulses on the fragments, thus suggesting the process to be violent. In Sekanina's model the splitting process is gentle. Sekanina gives a number of arguments that favor his model. One advantage is that the model requires only two free parameters: the time of splitting and the differential nongravitational force. In contrast, the old theory required four free parameters: the time of splitting and three initial differential velocity components. The Sekanina model has been applied to a number of split comets, including periodic Comet Biela, with very good results for the motion of one fragment relative to the other. The theory failed to predict the motion of the fragments of Comet Wirtanen.

Comet West (1975n) split into four fragments (see Figure 7.20) in early 1976. Using his theory of differential nongravitational effects, Sekanina has attempted to model the splitting process and subsequent relative motions. He has arbitrarily labeled the fragments *A, B, C,* and *D*, with *A* being the principal fragment; *C* was apparently very short-lived (March 12–25, 1976).

The probable course of events, according to Sekanina, was as follows. The parent nucleus broke into fragments *A* and *D* roughly a week before perihelion passage of the comet; then fragment *B* separated from either *A* or *D* about two days after perihelion. Fragment *C* next broke away from one of the other fragments (probably *A*) roughly a week to 10 days after perihelion. Thus, Comet West appears to have suffered three separate fragmentation processes within a week or so on either side of perihelion. The relative motions of the fragments ranged from about 0.25 to about 1.75 meters sec^{-1}. These conclusions are based on a number of multiparameter fits to the observed motions of the fragments. The fits with the smallest residuals were chosen for each pair of relative motions. The sequence of events Sekanina describes represents a best fit to the data and should not be considered definitive.

We have mentioned that we prefer an albedo for the nucleus of approximately 0.7. Delsemme and Rud (1973) have recently proposed a method (Chapter 5) for determining the albedo (and radius), and they find values for the visual albedo in the range 0.6 to 0.7. However, the method depends critically on the assumed nuclear material, and we defer this discussion until the evidence for the current model is summarized.

5 Model and origins of comets

A comet model

The specification of a comet model is critically dependent on the basic icy material or materials that comprise the bulk of the cometary nucleus, along with the micron-size dust that forms the type II tail and the larger dust responsible for meteor streams. The model we present is Whipple's icy-conglomerate model as updated by Delsemme and his colleagues to include clathrate hydrates. The dominant constituent of the nucleus is most likely water ice. To begin this chapter, we will present the evidence in favor of water-ice.

The vaporization of cometary snows is governed by the equation

$$F_0(1 - A_0) \frac{\cos \theta}{r^2} = (1 - A_1) \sigma T^4 + Z(T) \cdot L(T) \tag{5.1}$$

Here F_0 is the solar constant; A_0 is the albedo in the visual wavelength range ($\lambda \sim 5000$ Å); θ is the solar zenith angle on the nucleus; r is the heliocentric distance of the comet; A_1 is the albedo in the infrared region (15–30 μm) that dominates the thermal emission from the nucleus; σ is the constant in Stefan's law; T is the temperature; $Z(T)$ is the vaporization rate in molecules per centimeter squared-second; and $L(T)$ is the latent heat of vaporization in calories per mole. Solution of equation 5.1 yields both the temperature of the nucleus and the vaporization flux. Physically, equation 5.1 states that the solar radiant energy absorbed by the nucleus is disposed of by radiation back to space and/or by vaporization of water snows. When a balance between incident solar radiation and reradiation alone is assumed, the temperature is given by equation 2.10. Radiative processes alone determine the nuclear temperature at distances beyond approximately 4 AU. At smaller distances, the vaporization of water snows is an important energy loss for the nucleus, and it completely dominates the situation for distances less than 0.8 AU.

Equation 5.1 can be integrated over the surface of the nucleus to yield

$$F_0(1 - A_0) \frac{S}{r^2} = 4S(1 - A_1) \sigma T^4 + 4SZ(T)L(T) \tag{5.2}$$

where S is the cross section of the nucleus or πr_c^2. When the solar radiant energy is dissipated entirely by vaporization, equation 5.2

128

becomes

$$F_0(1 - A_0)\frac{S}{r^2} = 4SZL = QL \tag{5.3}$$

where Q is the rate of production of varporized molecules (molecules per second). If L is constant, Q varies as r^{-2}; this regime is shown in Figure 5.1 and is the straight line portion of the various curves in the log–log plot. When thermal radiation by the nucleus becomes important, the vaporization rate drops rapidly. Note that almost any substance besides water has $Q \propto r^{-2}$ in the inner solar system. The latent heat of vaporization of water does not vary strongly over the temperature range of interest; it decreases from 11,700 cal mol^{-1} at 150 K to 11,220 cal mol^{-1} at 250 K. The value of 11,480 cal mol^{-1} at 200 K is a good approximation for most comet work. Hence, as a comet approaches the sun in the regime described by equation 5.3, the increased

Figure 5.1. Vaporization rate Z for various snows as a function of heliocentric distance. Based on the steady state temperature of a rotating nucleus with albedo of 0.1. (Courtesy of A. Delsemme, University of Toledo)

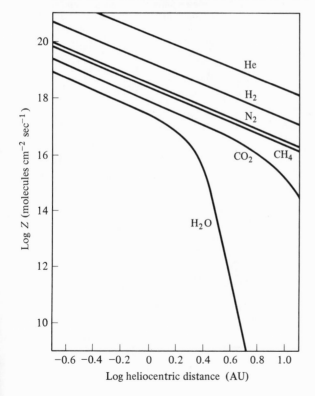

solar flux is balanced by an increased vaporization rate. However, as the comet nears the sun, the temperature of the nucleus increases very slowly (Figure 5.2). The temperature near 1 AU is approximately 200 K and varies as $r^{-0.12}$ at 4 AU and $r^{-0.06}$ at 0.3 AU.

Physically, the vastly increased vaporization rates required by equation 5.3 when a comet approaches the sun are explained as follows. The crucial quantity is the pressure P_E of the vaporized gas in equilibrium with vaporizing snows. This pressure is given by

$$P_E = N_E kT \tag{5.4}$$

and is a function of temperature. Here N_E is the gas density at equilibrium (molecules cm^{-3}). The vapor in equilibrium is said to be saturated. The equilibrium is maintained by a balance between the vaporizing flux (Z^+) and the condensing flux (Z^-). On the kinetic theory model of condensation, all molecules that collide with the snow surface condense. Hence, Z^- can be calculated from the density in the vapor, the mean speed of the molecules, and geometrical factors. Of course, this

Figure 5.2. Temperature at the subsolar point on a cometary nucleus composed of water snows as a function of heliocentric distance. Labels along the curve indicate the local rate of variation. Note that the value of $T \propto r^{-1/2}$, the normal value for nonsublimating materials, is reached only for $r = 10$ AU. (Data from A. Delsemme, 1966, *Mém. Soc. R. Sci. Liège 37*:69–76)

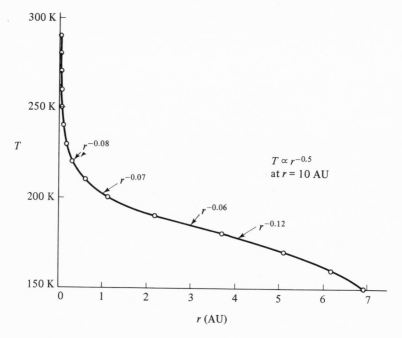

quantity is just the number of collisions of gas molecules with the surface or

$$Z^- \text{ (equilibrium)} = \tfrac{1}{4}N_E\langle V\rangle \tag{5.5}$$

where the mean speed for a Maxwellian distribution is

$$\langle V\rangle = \left(\frac{8kT}{\pi m}\right)^{1/2} \tag{5.6}$$

At equilibrium the vaporizing flux equals the condensing flux from the saturated vapor. But if the saturated vapor is replaced by a vacuum, the vaporizing flux does not change and is given by equation 5.5. Expressing the density at equilibrium in terms of the vapor pressure (equation 5.4), we can write the equation for Z as

$$Z = \frac{P_E}{(2\pi mkT)^{1/2}} \tag{5.7}$$

The vapor pressure for ice can be obtained from standard chemical tables. The quantity exhibits a steep variation with temperature. Empirically, near 200 K, a 20 K or 10% increase in temperature increases P_E (and hence roughly Z) by a factor of 20. Thus, a minor change in temperature produces the energy balance implied by equation 5.3.

Theoretically, the variation in vapor pressure is given by standard thermodynamic expressions. A phase change that takes place at constant temperature and pressure is described by Clapeyron's equation:

$$\frac{dP_E}{dT} = \frac{L}{T(V_g - V_s)} \tag{5.8}$$

where L is the latent heat of vaporization and V_g and V_s are the volumes of the gas and solid, respectively. In the present situation $V_g \gg V_s$, so we can ignore the latter quantity. If we express V_g in terms of the ideal gas law $V_g = RT/P_E$, we can express equation 5.8 in the form

$$d \ln P_E = - \frac{L}{R} \, d(1/T) \tag{5.9}$$

Because L is approximately constant, equation 5.9 can be integrated to yield

$$P_E \propto e^{-L/RT} \tag{5.10}$$

For temperatures around 200 K, the exponential term becomes, for water, $e^{-5740/T}$, showing the extreme temperature sensitivity of P_E.

This picture is entirely consistent with the evidence obtained from studies of nongravitational forces in comets. Refer to our previous discussion of the nongravitational forces in Chapter 3. Water is the prime candidate for the substance controlling the vaporization rate because other possible dominant constituents (such as CO, CO_2, CH_4, and NH_3) have $Q \propto r^{-2}$ to such large distances (Figure 5.1) that the orbital residuals for P/Schwassmann–Wachmann 2 and other comets are entirely unacceptable. The exact distance at which the vaporization flux begins to fall off much faster than the r^{-2} law depends on the visual and infrared albedos. Sample curves showing the vaporization rate versus r with different visual and infrared albedos are shown in Figure 5.3. This distance r_0 (defined by equation 3.14) can be written approximately as

$$r_0 \approx 2.8 \left[\frac{1 - A_0}{1 - A_1}\right]^{1/2} \quad (AU) \tag{5.11}$$

where A_0 and A_1 are the visual and infrared albedos, respectively. For equal visual and infrared albedos, r_0 is 2.8 AU and the variation of r_0 with albedos is shown in Figure 5.4. For our assumed model of water snows, the distance r_0 corresponds to the distance at which the energy

Figure 5.3. Vaporization flux of water snow as a function of heliocentric distance. Each curve is characterized by a visual albedo A_0 and an infrared albedo A_1, written as (A_0, A_1). (Courtesy of B. G. Marsden, Z. Sekanina, and D. K. Yeomans)

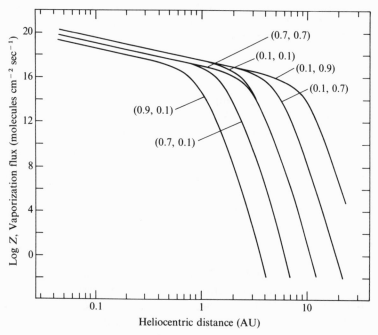

lost in reradiation is approximately 43 times the energy lost in vaporization. This small balance from reradiation is important because even a
small decrease in temperature produces a large decrease in vaporization rate. Hence, Q falls off rapidly beyond r_0.

The appearance of molecular emissions near 3 AU is readily explained by a nucleus in which minor constituents are trapped in the
cavities of clathrate snows. A variation in the distance at which molecular emissions appear occurs in part due to a variation in albedos. (The
sequence in which molecular emissions occur depends on excitation
and ionization by solar radiation.) The rate of release of coma gases
such as CN or C_2 is determined by the vaporization rate of water ice,
specifically, the sublimation rate of the icy lattice of the clathrates.
Delsemme has shown that adsorption on water snows leads *ther-
modynamically* to the formation of clathrate hydrates of gas (see Chapter 4). If all available cavities are filled, the adsorbed gas comprises 1
molecule to $5\frac{3}{4}$ water molecules, or 17% of the icy material. If the
amount of gas exceeds this limit, it will condense without adsorption.
Because most frozen gases are more volatile than water ice, they will
be the first substances available for evaporation. Such excess gases
could produce comas at large heliocentric distances on a comet's first
perihelion passage, and their loss probably explains the dimming of
most new comets after the first perihelion passage. It appears then that
the vaporization rate of nuclear material is completely controlled by
water snows for the second passage onward.

This assumption allows the albedos A_0 and cross-sectional area S
(radii) of cometary nuclei to be calculated. Observations of comets at
large heliocentric distances determine A_0S with relatively small con-
tamination from the coma. At small heliocentric distances, determination of the total vaporization rate Q determines $(1 - A_0)S$ (equation
5.3). These two independent observations were combined by Delsemme and Rud (1973) to yield the results shown in Table 5.1; they are
most encouraging and entirely consistent with our other knowledge of
comets. An attempt to apply this method to Comet Encke yields incon-
sistent results, which indicates that the physical situation is fundamen-
tally different. Perhaps water ice covers only a small fraction of the
nuclear surface.

Table 5.1. *Albedos and radii of cometary nuclei*

Comet	A_0	R(km)
Tago–Sato–Kosaka	0.63 ± 0.13	2.20 ± 0.27
Bennett	0.66 ± 0.13	3.76 ± 0.46

A water-ice nucleus easily explains the origin of the three most abundant substances found in comets, namely, H, OH, and O. Abundances of nearly all other observed species are two orders of magnitude lower. A conceivable exception is CO and possibly CO_2; a large CO abundance was inferred from observations of Comet Kohoutek (Chapter 7) and would naturally explain the importance of CO^+ in comet tails. However, CO_2 could be the parent of CO and by dissociation would also produce the proper amount of O in the ^1D state, required to explain the brightness of the forbidden red line in most comets by the process

$$CO_2 + h\nu \rightarrow CO + O(^1D)$$

The rate of production of H – in the comets where this quantity has been inferred – is also consistent with the water ice model. The injec-

Figure 5.4. The distance r_0 in AU for water snow, as a function of the visual albedo A_0 and the infrared albedo A_1. (Courtesy of B. G. Marsden, Z. Sekanina, and D. K. Yeomans)

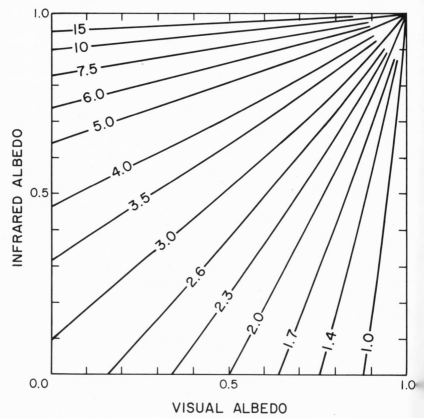

tion velocity for the atoms in the hydrogen cloud is about 8 km sec^{-1} (Chapter 4), and this is consistent with the outer H atoms arising from the photodissociation of OH. The brightness variation of the hydrogen and hydroxyl clouds of Comet Tago–Sato–Kosaka was measured by Code, Houck, and Lillie (1972) to vary as r^{-6}. This agrees with the icy-nucleus model as long as the distinction between the variation in the monochromatic brightness of the head and the variation in the production rate Q is kept in mind. Delsemme (1973b) has shown that the luminosity variation as seen through a diaphragm of specific size is

$$L(r) \propto r^{-n-\alpha} \qquad (5.12)$$

where n is the exponent in the production law and α is the correction due to finite diaphragm size. For small diaphragms, $\alpha = 3.5$, and this result coupled with $n \approx 2.5$ for the production law yields the observed r^{-6} brightness variation. To the accuracy that these parameters are known, $n \approx 2$ is equivalent to $n \approx 2.5$.

If the visual albedo A_0 is taken to be about 0.6 (see above), the relation between the infrared albedo A_1 and the production rate Q can be calculated (see equation 5.2). The results are shown in Table 5.2. If n is ultimately established to be 2.5 and A_0 is taken to be 0.6, the tabulated figures imply an infrared albedo of about 0.4. Note that most nonmetallic substances have infrared albedos less than 0.5. Then equation 5.11 or Figure 5.4 determines an r_0 of about 2.5 AU. Thus we have working values for the size, albedos, and water vaporization rates from cometary nuclei. All are consistent with facts associated with or derived from the hydrogen or hydroxyl clouds. They are also consistent with the gas production rates necessary to drag the proper amount of dust off nuclei and to give the dust the necessary speed required to explain type II tails in the Finson–Probstein model of dust tails. The

Table 5.2. *Variation of exponent n in the vaporization rate of water snows with heliocentric distance,* $Q \propto r^{-n}$ *for 0.8 AU $\leq r \leq$ 1.0 AU*

Infrared albedo (A_1)	Visual albedo (A_0)		
	0.5	0.6	0.7
0.2	2.69	2.88	3.25
0.4	2.50	2.63	2.83
0.6	2.32	2.40	2.49

Source: After Delsemme (1973).

first results for Comet Arend–Roland indicated this result and were established in detail for Comet Bennett by Sekanina and Miller (1973). The vaporization flux required (at 1 AU) is 4×10^{17} molecules $(cm^2\text{-}sec)^{-1}$, a value most compatible with the vaporization of water snows. A summary schematic of the cometary nucleus is given in Figure 5.5.

Delsemme and Wenger (1970) have conducted laboratory studies of the properties of clathrate-hydrate snows. They have the appearance of a peculiar powdery snow composed of icy grains 0.1–1.0 mm in diameter. When a cometary environment is simulated by a vacuum, the evaporating gases strip grains from the main body of the snow. In comets the nucleus should be surrounded by a halo of such icy grains carried away by the escaping gas. Escaping grains are retarded by the gravitational attraction of the nucleus and vaporized by solar radiation. It is relatively simple to calculate the radius of the icy grain halo for different grain sizes and typical parameters (see Figure 5.6). At 1 AU and for grain sizes between 0.1 and 1.0 mm, the radius of the halo is in the range 1×10^3 to 3×10^3 km. Values in the range 10^3 to 10^5 km are considered possible. The size of the icy grain halo within the (gravita-

Figure 5.5. Schematic representation of the cometary nucleus. (Courtesy of D. W. Hughes, University of Sheffield)

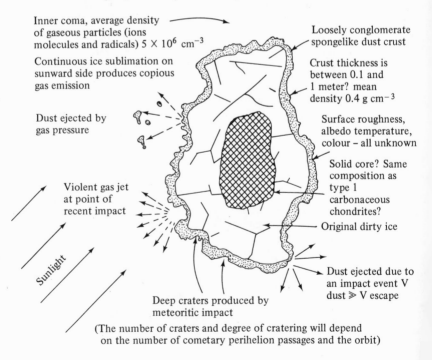

Inner coma, average density of gaseous particles (ions molecules and radicals) 5×10^6 cm^{-3}

Loosely conglomerate spongelike dust crust

Continuous ice sublimation on sunward side produces copious gas emission

Crust thickness is between 0.1 and 1 meter? mean density 0.4 g cm^{-3}

Dust ejected by gas pressure

Surface roughness, albedo temperature, colour – all unknown

Solid core? Same composition as type 1 carbonaceous chondrites?

Violent gas jet at point of recent impact

Original dirty ice

Sunlight

Dust ejected due to an impact event V dust ≫ V escape

Deep craters produced by meteoritic impact

(The number of craters and degree of cratering will depend on the number of cometary perihelion passages and the orbit)

tional) cutoff decreases with decreasing heliocentric distance; as shown in Figure 5.6, the halo radius is approximately proportional to r. The density cutoff at the outer part of the icy grain halo is quite sharp. Delsemme and Miller (1970) have calculated the brightness profile of continuum light reflected from the halo. The false or photometric nucleus could result from the halo. In addition, the halo may help explain optical depth effects observed near the nucleus of Comet Burnham.

A most important application of the icy grain halo may be in understanding the lifetime and distribution of the parent molecules for coma constituents such as C_2 and CN. Before studies of Comet Kohoutek were made (see Chapter 7), not one parent molecule candidate responsible for minor constituents such as CN had been observed. The lifetime of possible parent molecules would be a valuable clue to the correct identification. Potter and Del Duca (1964) used absorption cross sections from laboratory data to calculate the lifetimes of numerous possible parent molecules to dissociation by solar ultraviolet radiation. Most of the calculated lifetimes are too large to explain the observations. For example, studies of brightness outbursts give a lifetime of

Figure 5.6. Extent of icy grain halos built up with grains of a single size, as a function of heliocentric distance. The size of the grains as observed in the laboratory lies between the two solid lines. The envelope for all halos is marked by the horizontal lines labeled $T = 200$ K and $T = 50$ K, corresponding to the assumed temperature of the nucleus. (Courtesy of A. H. Delsemme, University of Toledo)

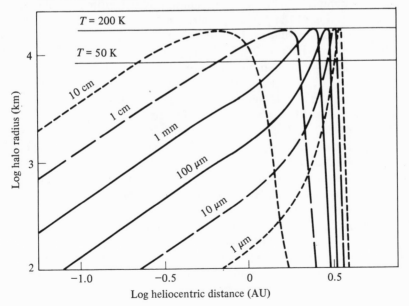

about 1 hr for the CN parent, and other studies are compatible with this short lifetime. The shortest lifetime for a possible parent of CN was found for monocyanoacetylene (CNC_2H_2), which dissociates in approximately 18 hr. This is at least an order of magnitude too large to explain the production of CN in comets. Similar discrepancies exist in the lifetimes of the parent molecules for CH and C_2. Molecules not yet considered may eventually prove to be the missing parents, but no good possibilities are in sight. Highly unstable complex radicals could be stored in the icy matrix of the nuclear ices, and they might provide the needed parent.

Icy grains moving away from the nucleus could store the radicals themselves (such as CN, C_2, and CH) in the icy matrix, and these would be released as the grains vaporize. Alternatively, the lifetime of the parent molecule must be very short. Delsemme has suggested that an icy grain halo composed of relatively large particles (size ~ 1 cm) could be the source for most minor constituents observed in the coma. Vaporizing halos of approximately 2.5×10^4 km at 1 AU could account for the observed CN and C_2 distributions in comets Burnham (1960 II) and Bennett (1970 II). The heliocentric variation of the halo sizes for Comet Bennett seems closer to the proportional law ($r_H \propto r$) than to other variations. So far, then, the evidence is favorable but not conclusive concerning the importance of the halo of vaporizing grains in providing an effective parent molecule for minor constituents.

The picture summarized above and in the previous chapter is reasonably complete in broad outline, and we can legitimately claim some understanding of the nucleus, hydrogen cloud, coma, and normal dust tails. The tails of distant comets (such as Comet Baade 1955 VI and Comet Haro–Chavira 1956 I) also appear to be explicable on the basis of the Finson–Probstein modification of the Bessel–Bredichin mechanical theory. The two basic problems are their orientation (making $\approx 45°$ angle with the radius vector) and their existence itself. Sekanina has pointed out that there is a tendency for the synchrones of the tail to lie on top of each other when the comet is approaching the sun from a large distance. The tails of the distant comets have orientations consistent with emission some 200 to 2000 days before perihelion (corresponding to r in the range of 5 to 15 AU). The size of the particles must be 100 μm or larger. This size range is unusual in cometary physics except for the icy grains stripped from the snows of the clathrate hydrates. At heliocentric distances of 4 AU or greater, the icy grains would have a very long lifetime. Hence, all that the distant tails require for their existence is that the cometary nucleus contain an effective reservoir of substances more volatile than water snow, in order to provide the momentum needed to strip the clathrate grains from the nucleus. It is reasonable to expect such substances in comets that remain at large distances from the sun or that are approaching the sun for the first time.

Our basic understanding of most parts of a comet does not yet carry over to the plasma tail and associated phenomena. Part of the uncertainty of the physics of the plasma tail arises because many important features involve the interaction between the comet and the solar wind. In other words, the solar particle emission and its properties determine the plasma properties of the comet. Another part of the uncertainty arises because the solar wind interaction is thought to produce phenomena that as yet have not been observed. We believe the solar wind flow is disrupted over dimensions comparable to the size of the hydrogen cloud, but most of our data refer to the central, visible plasma tail. In addition, the flow in the solar wind is supersonic and super-Alfvénic. Hence, shock fronts may be a feature of the flow models in a way generally analogous to the earth's bow shock in the solar wind–geomagnetic field interaction.

Among the first systematic investigations of this problem was the study by Biermann, Brosowski, and Schmidt (1967); a modern version of their overall model is schematically shown in Figure 5.7. Axial symmetry is assumed, and the results obtained apply to the area of the

Figure 5.7. Schematic representation of the interaction of the solar wind with a comet. In addition to the features discussed here, there may be an inner shock in the flow of the coma gas. (Courtesy of J. C. Brandt and D. A. Mendis)

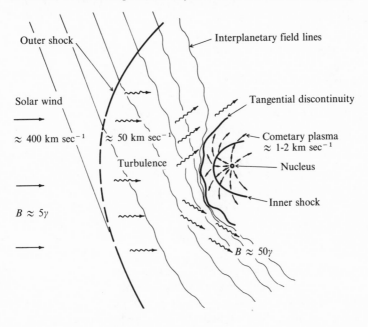

sun–comet line. The cometary CO is emitted isotropically and ionized by photoionization only. When ionized, the CO^+ ions are rapidly accelerated to the local flow speed of the plasma. Thus, the equations that describe the flow of the plasma contain a source term to account for the mass of the CO^+ ions added by ionization; in other words, mass is not conserved. Supersonic flows are very sensitive to the addition of heavier molecules, and a detached shock develops when the plasma contains 1% CO^+ ions by number. The shock is calculated to be on the order of 10^6 km from the nucleus. This distance is generally compatible with flow speeds in the coma ~ 1 km sec^{-1} and a photoionization lifetime $\sim 10^6$ sec. A stagnation point some 10^5 km from the nucleus is also found, and this fact is interpreted in terms of a contact discontinuity that separates the pure cometary plasma from the mixed cometary and solar wind plasma.

If the distance to the contact discontinuity is interpreted as corresponding to the sunward extension of the CO^+ envelopes, the calculated distance is an order of magnitude larger than the observed value $\sim 10^4$ km observed in comets Morehouse and Halley. Biermann, Brosowski, and Schmidt themselves noted the simplifications in their model. The efficiency of ionization could be enhanced by charge exchange with protons, ionization by energetic electrons, and chemical reactions near the nucleus. Extension of the earlier work to the three-dimensional case with axial symmetry by Brosowski and Wegmann (1972) for the same ionization parameters confirms the approximate values for the distances to the contact surface and the shock front. Hence, the problem of the production of the CO^+ close to the sunward side of the nucleus (within 500 to 1000 km) remains a puzzle in the understanding of plasma tails. At the least, it seems to imply that un-ionized CO is an abundant constituent of the cometary atmosphere, perhaps comparable to the abundance of OH. We return to the discussion of CO in considering the origin of comets and new results from Comet Kohoutek. Delsemme (1975) argues that CO^+ could come from CO_2 by a charge exchange reaction, in which case the presence of the molecular ion does not imply a large abundance of CO.

Brosowski and Wegmann's model does (at least qualitatively) account for the flow pattern around a comet. The flow lines are shown in Figure 5.8; note that the contact surface is assumed to be hemispherical on the solar side and cylindrical on the antisolar side. This flow pattern permits a calculation of the motion of a CO^+ filament, as shown in Figure 5.9. The filament remains approximately a straight line, lengthens, and turns toward the contact surface. This behavior is entirely consistent with our knowledge of filament or tail ray evolution. The filaments themselves could arise from structures in the solar wind plasma. Hence, the flow pattern illustrated in Figure 5.8, a flow essentially parallel to the tail axis and with a lateral gradient ~10 km sec⁻¹ per 10⁵ km, is consistent with our empirical data. On this picture, the flow speed reaches the speed of the surrounding solar wind at distances ~10⁶ km from the axis. We conclude by noting that the disruption of the solar wind flow by the comet can influence the lifetime of neutral hydrogen to charge exchange by solar wind protons (see Chapter 6).

We conclude our discussion of the comet model with several schematic diagrams (Figures 5.10, 5.11, and 5.12) showing the principal features of the comet model and summarizing the discussion given in Chapters 4 and 5.

The origin of comets

In discussing the origin of comets, we assume that all comets have an essentially common physical origin. A concomitant assumption

Figure 5.8. Flow lines on the Brosowski–Wegmann model. (Courtesy of L. Biermann, Max-Planck-Institut für Physik und Astrophysik)

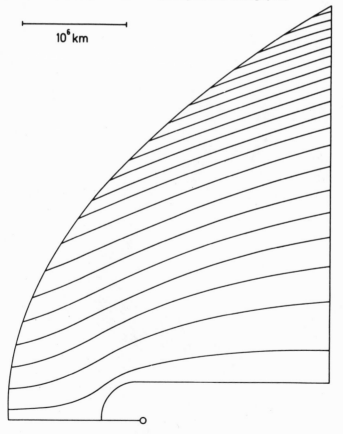

is that the short-period comets are a natural consequence of repeated passages through the inner solar system by the long-period or nearly parabolic comets. We will address the physical process of short-period comet production later in this subsection. Subsequent developments may require separate processes for the origins of the short-period and long-period comets; but in the interest of simplicity, we should not consider that more complex possibility unless it is clearly indicated by very persuasive evidence.

Hypotheses concerning the origin of comets must be compatible with two basic facts: (1) the lack of substantial numbers of comets with hyperbolic orbits indicates an origin within the solar system or at least in a system with exactly the same space motion as the solar system; and (2) the composition of comets shows similarities with the composition of interstellar clouds. At the very least, the composition of comets

Figure 5.9. Motion of a filament on the Brosowski–Wegmann model. (Courtesy of L. Biermann, Max-Planck-Institut für Physik und Astrophysik)

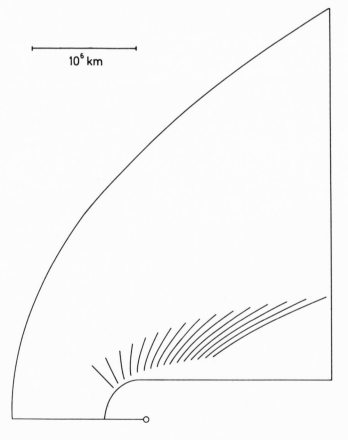

10^6 km

Figure 5.10. Schematic of principal gaseous features of a typical comet on a logarithmic scale. (*Report of the Comet Science Working Group*, NASA–Jet Propulsion Laboratory, 1979)

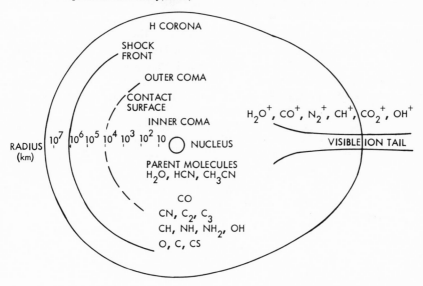

Figure 5.11. Sketch of principal dusty features of a typical comet on a linear scale, unit of 10^4 km. (*Report of the Comet Science Working Group*, NASA–Jet Propulsion Laboratory, 1979)

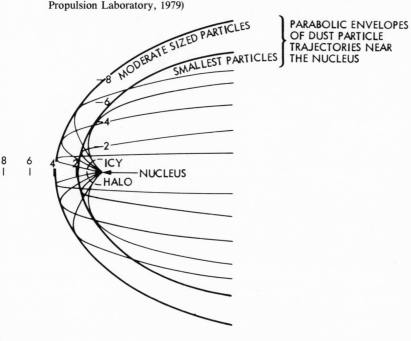

is consistent with condensation from interstellar clouds (see Table 5.3). In this subsection, our working hypothesis for the origin of comets is that they are natural products of condensation in the contracting solar nebula. Near the sun this condensation process led to the formation of the terrestrial planets and, at larger distances, the Jovian planets. New comets, thus, may be essentially unchanged products that are representative of conditions in the solar nebula. Hence, new comets may be a source of information for the early history of the solar system when it was in the process of formation from collapsing interstellar material.

Figure 5.12. Features and processes involved in the interaction of a comet with sunlight and the solar wind. (*Report of the Comet Science Working Group,* NASA–Jet Propulsion Laboratory, 1979)

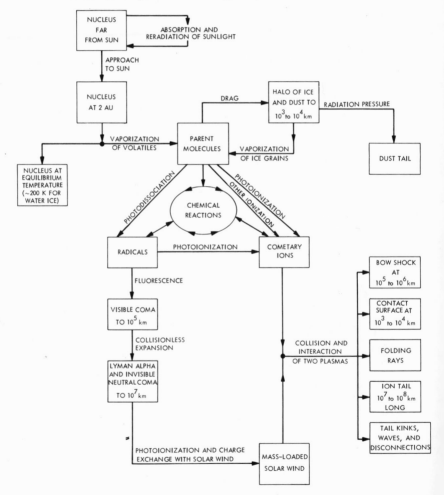

The remainder of this section is an attempt to develop this working hypothesis and to fill in details wherever possible. Note, however, that the hypothesis of origin in the inner solar system has had a recent interesting revival.

First, we discuss the evidence based on orbital considerations. There is currently no serious question as to the existence of a peak in the frequency of $1/a$ values, with the peak falling between 10^{-4} and 10^{-5}. This peak has been identified with the outer fringe of the Oort cloud of new comets. It shows up in numerous discussions, including Oort's original discussion, and extending to recent discussions such as Delsemme's (see Figure 5.13). We do not discuss immediately the origin of the Oort cloud with dimensions of 10^4 to 10^5 AU at least partly because the stellar perturbations that send new comets toward the inner solar system also operate to make the distribution of cometary aphelia isotropic. In other words, because of stellar perturbations the comets in the Oort cloud should retain no orbital memory of their origin. Naturally, this situation complicates discussions of the origin of comets.

We address the following question: Can the presently observed number of short-period comets be explained through capture by Jupiter from the population or near-parabolic comets? The numerical (Monte-Carlo) treatment by Everhart (1972) seems to explain the properties of the short-period comets (periods less than 13 years), with the possible exception of their absolute rate of capture. Indeed, Joss (1973) has discussed the problem using Everhart's data, and he concluded that the observed number of short-period comets is approximately 4×10^4

Table 5.3. *Chemical constituents of comets and the interstellar medium*

Products observed in comets	Possible precursors observed in interstellar space	Possible parent molecules observed in comets
H	Many observed molecules	Many observed molecules
O, OH, OH^+	$H_2O, H_2C=O, H_3C-OH$	H_2O, H_2O^+
CO^+	$CO, H_2C=O, H_3C-OH, NH=C=O$	
CH, CH^+	$H_2C=O, H_3C-OH, HC=C-C\equiv N$	$H_3C-C\equiv N$
C_2, C_3	$HC\equiv C-CH_3, HC\equiv C-C\equiv N$	$\begin{cases} CH_3-C\equiv C-H \\ CH_3-C\equiv C-CN \end{cases}$
CN	$H-C\equiv N, H-N=C, HC\equiv C-C\equiv N$	$H-C\equiv N$
NH, NH_2	$NH_3, H_2N-CH=O, HN-C=O$	
N_2^+, CO_2^+	Other precursors or mechanisms involving collisions	

Source: Based on a table prepared by A. H. Delsemme (1977).

times too large. Delsemme (1973) has recently rediscussed the problem, and we summarize his discussion here.

In a steady state the rate of destruction of the short-period comets is equal to the rate of capture by Jupiter. This situation can be expressed mathematically as

$$\frac{N_s}{\langle P \rangle l_0} = \langle f'_c \rangle N'_0 \tag{5.13}$$

Here, N_s is the number of short-period comets; $\langle P \rangle$ is the mean period of the short-period comets; l_0 is the number of revolutions of the short-period comets before disintegration; $\langle f'_c \rangle$ is the mean capture probability for a comet passing within Jupiter's zone of capture; and N'_0 is the number of comets per year passing perihelion within the capture zone. The left-hand side of equation 5.13 is the rate of destruction of short-period comets, and the right-hand side is the injection rate due to Jupiter's influence.

Figure 5.13. Normalized distribution of the semimajor axes for the observed passages of all periodic comets. N is the number that pass perihelion in a Δq of 1 AU, per $1/a = 1\ \mathrm{AU}^{-1}$, per century. A is the extra number of "new" comets in Oort's peak. Segment AB of the histogram is the distribution of the original orbits of long-period comets. BD is the distibution of the intermediate-period orbits, down to Jupiter's 2:1 resonance. C shows the peak of short-period comets of Jupiter's family. Other curves are discussed in the text. (Courtesy of A. H. Delsemme, University of Toledo)

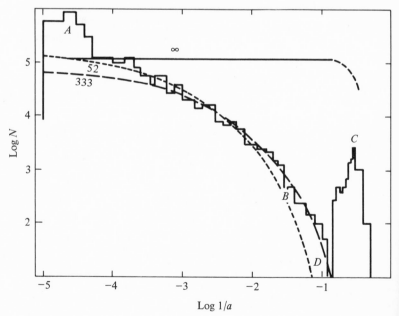

Estimating N_0' is most difficult. Comets with perihelia in the range 4 to 6 AU have disintegration lifetimes greater than 5×10^9 years because vaporization due to solar heating is almost negligible at these distances. The characteristic time for orbital diffusion is less than 5×10^6 years. As van Woerkom (1948) has shown, this situation leads to a distribution in which the number of comets that pass perihelion each year is constant in constant intervals of $1/a$. This distribution corresponds to the horizontal line labeled ∞ in Figure 5.13. The constant value is determined from the essentially flat distribution of observed comets near $1/a = 10^{-4}$, and this determines the number of comets passing perihelion between 4 and 6 AU per year. The capture zone includes only comets with sufficiently low inclinations; allowances are also made for incompleteness and the fact that most of the comets of interest (largely intermediate-period comets with $a < 30$ AU) are somewhat concentrated in the plane of Jupiter's orbit. The result is an N_0' corresponding to 120 passages per year.

The capture rate is based on Everhart's value of 2.5×10^{-4} per passage for parabolic orbits. The capture probability should be higher for intermediate-period comets, and Delsemme adopts a capture probability of 5×10^{-4} per passage. We stress that we are not referring to capture by a single encounter, but rather capture as a cumulative effect.

The mean period of the short-period comets ($P < 13$ yr) is readily found to be 7 years. The average number of perihelion passages l_0 can be deduced from the distribution in $1/a$ of comets that have penetrated the inner solar system and been subjected to vaporization. The curves marked 52 and 333 in Figure 5.13 are the theoretical curves on Oort's models, with half times for decay of 52 and 333 passages, respectively. Delsemme adopts $l_0 = 200 \pm 100$ passages for the short-period comets.

We may collect our results and rewrite equation 5.13 as

$$N_s = 84 \left(\frac{\langle f_c' \rangle}{5 \times 10^{-4}} \right) \left(\frac{N_0'}{120 \text{ yr}^{-1}} \right) \left(\frac{l_0}{200} \right) \left(\frac{\langle P \rangle}{7 \text{ yr}} \right) \tag{5.14}$$

Hence, the values of the various parameters chosen yield a steady-state value of 84 short-period comets. The observational value is 73 short-period comets. Delsemme feels that the estimated value could easily range from 20 to 200, but the agreement is encouraging. Nevertheless, this calculation should be regarded primarily as indicative of the problem.

Delsemme's solution requires some 1000 to 3000 intermediate-period comets to pass perihelion between 4 and 6 AU per year. The large distance and lack of activity make these comets practically unobservable. The only known example of this population is the giant Comet P/Schwassmann-Wachmann I ($q = 5.54$ AU, $P = 16.1$ yr).

Although the hypothesis that the short-period comets can be produced from the population of long-period (nearly parabolic) comets is not proven, it is plausible on the basis of the evidence outlined above. The nearly parabolic comets come from a cloud with dimensions $\sim 10^4$ to 10^5 AU, that is, the Oort cloud. These dimensions are approaching the distances to the nearest stars. Because the orbital inclinations of long-period comets are distributed roughly at random, it is likely that the Oort cloud is spherical. Perturbations due to passing stars have several important effects on the Oort cloud. First, the size of the cloud is limited to approximately 10^5 AU. Second, the stellar perturbations continually send new comets into the inner solar system where we observe them. If these perturbations did not occur, only comets in the cloud with very small transverse speeds could reach the inner solar system. Such comets would comprise a very small fraction of the total cloud, and Oort has estimated that this supply would be exhausted in a few million years. The total number of comets in the reservoir can be estimated from the number of new comets seen yearly. Oort found $\sim 10^{11}$ comets with a total mass in the range 10^{-1} to 10^{-2} $M_\oplus$. The same stellar perturbations that send comets toward the sun also send comets away from the sun, and these are lost to interstellar space. Third, the stellar perturbations tend to randomize the distribution of orbits in the Oort cloud. This fact means that clues concerning the origin of comets (such as an initial distribution possibly confined to the ecliptic plane) may have been erased by stellar perturbations if sufficient time has passed. Hence, orbital statistics might be of limited value in probing the site or mechanism of the ultimate origin of comets. Thus, stellar perturbations may be as important as planetary perturbations and disintegration in determining the statistics of cometary orbits.

Evidence from the physics of condensation within the solar nebula constitutes the principal means of distinguishing between alternate theories of the physical origin of comets. In his 1950 paper, Oort noted that the formation of comets at the great distances from the sun characteristic of the reservoir was very unlikely because of the very low (essentially interstellar) densities involved. Hence, he suggested that the comets were probably formed in the same general region as the planets and were subsequently expelled by planetary perturbations. The process would be diffusion of the same kind as discussed above, except that the direction would be outward. Many authors have noted that this process is intrinsically inefficient because most comets expelled from the inner solar system would leave the solar system, whereas only a small number would achieve orbits with the necessary large semimajor axes. To some, Oort's picture is aesthetically unsatisfactory because the major planets are used both to eject comets initially and to capture short-period comets currently. Whatever the difficulties of Oort's scheme, it supplies a complete picture of the origin of comets with some attention to the relevant physics.

In 1950 Oort addressed the problem of the ultimate origin of comets by postulating a common origin with the asteroids – namely, both are fragments of a broken-up planet. This idea was tenable when the cometary nucleus was thought to be made of stone; the icy-conglomerate nucleus probably requires another type of origin. Indeed, the breakup of a Jovian planet has recently been suggested and perhaps does not have this difficulty. The current consensus view of cometary origin tends toward condensation processes associated with the formation of the sun and the planets. Although we will pursue this view, we note the many divergent views on this subject; see, for example, Chebotarev, Kazimirchaka-Polonskaya, and Marsden (1972).

Modern research on condensation processes in the collapsing solar nebula has been active, and some results relating to the formation of comets are available. Naturally, there is as yet no consensus concerning details, and the following discussion is intended only as an illustration. If we consider the condensation processes as taking place in chemical equilibrium, the temperature–pressure relationships for condensation of various substances are as shown in Figure 5.14. The adiabat indicates a temperature–pressure relation in the nebula at an early stage in its history. The formation conditions along the adiabat for the various planets are marked by the relevant planetary symbol; these positions are calculated from Cameron's (1973) model of the solar nebula. Clearly, Figure 5.14 indicates condensation of higher-density material necessary for formation of terrestrial planets near the sun and condensation of lower-density material necessary for formation of the Jovian planets away from the sun. The simple chemical equilibrium model can reproduce the general features observed in the densities of the planets; rapid accretion processes (that would isolate material from further chemical reactions with the gas) may not be too important on the large scale.[1] On this model, we crudely expect the condensation of ice and the formation of comets to occur at temperatures ≈ 100 K, that is, roughly in the region between Saturn and Neptune. As the temperature decreases, solid hydrates are formed.

Specifically, Lewis (1972a,b) notes the following results for an ideal situation in which chemical equilibrium holds down to a certain temperature and accretion is then allowed to occur: (1) accretion near 150 K produces water ice with rocky material; (2) accretion near 100 K produces the solid hydrate of ammonia ($NH_3 \cdot H_2O$) with water ice and rocky material; (3) accretion near 50 K produces ammonia and methane hydrates. At the latter temperature, water ice has been converted to the hydrates mentioned. In summary, accretion of material in chemical equilibrium at temperatures ≈100 K is compatible with the Whipple–Delsemme picture of the cometary nucleus. Accretion at temperatures lower than roughly 50 K are not required by our present knowledge of cometary chemistry.

Thus, we can describe a plausible scenario for the formation of cometlike bodies of ice in the solar system between Saturn and Neptune, that is, at heliocentric distances approximately in the range 10 to 30 AU. However, we still have the problem (discussed above) of injecting these comets into the Oort cloud; presumably, this would occur as the result of close encounters with Jupiter or Saturn. Cameron again has

Figure 5.14. Major features of the chemistry of solar material. Condensation curves for representative material are shown, as well as an adiabat for the early solar nebula. See text for discussion. (Courtesy of J. S. Lewis, Massachusetts Institute of Technology)

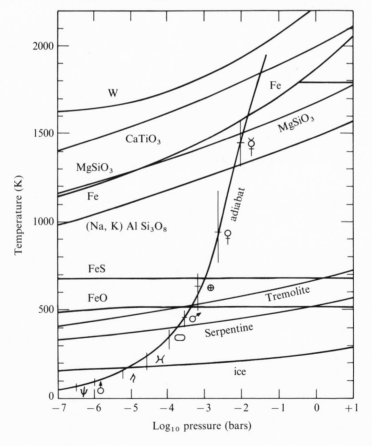

noted the unsatisfactory nature of this scheme and has discussed the formation of comets on the basis of Cameron and Pine's models of the primitive solar nebula. Basically, Cameron concludes that the angular momentum of the early solar nebula originated in random turbulence. Hence, it is plausible that the early solar nebula could have possessed satellite nebulae or "subdisks" in highly elliptical orbits. Cameron suggests that these were the sites of comet formation. Even a small number of subdisks would be sufficient to form a reservoir of comets. The velocity dispersion necessary for the observed spherical distribution would result from disintegration of the subdisks and stellar perturbations. Further, because the cometary accumulation takes place in widely varying physical circumstances, large differences in the physical characteristics of comets would be expected.

We have described one plausible scenario for the origin of comets that does not grossly violate known laws of physics or known facts about comets. The reader should regard this picture as indicative of the problems that remain to be solved. Cameron (1973) has described the general picture as follows: "The physical discussion has been quite crude, being carried out by 'back of the envelop' methods, although a large number of envelopes were needed. Thus the calculations are quite exploratory." This area is developing rapidly, and the student should refer to the current literature for the latest ideas.

One central theme of the picture presented is a feature of most schemes for the origin of comets. They are pristine, primitive bodies, probably remnants of the solar system's formation. To point out the many facets of this theme, we can do no better than to quote from the August 1979 *Report of the Comet Science Working Group* (p. 1):

1. Comets may contain a variety of presolar system, interstellar grains, which have so far been unavailable for study.
2. Comets probably retain evidence of the chemical and physical conditions under which they and the planets formed, especially about the processes of condensation, agglomeration, and mixing that took place.
3. Comets may have been a major source of the atmospheres of the terrestrial planets.
4. Comets may have provided the Earth with the organic molecules necessary for the evolution of life.
5. Comets are a source of meteors and interplanetary dust and possibly of some Apollo/Amor asteroids.

These topics, as well as the physical processes that produce the spectacular displays associated with comets, justify the intense, widespread interest in comets. This interest may culminate with in situ studies of comets in the 1980s as discussed in Chapter 8.

The origin of comets: a recent view

The last few years have seen a revival of the Lagrangian or planetary hypothesis of the origin of comets. The impetus for the revival has been dynamic calculations by the Canadian astronomer M. W. Ovenden. He has carried out numerical experiments on the motion of multiple-planet N-point-mass systems to show that the time a system spends in a given configuration is inversely related to the magnitude of the mutual interactions between the planets in that configuration. Ovenden has generalized this finding to a *principle of least interaction action*. The principle predicts that the solar system will tend toward a Bode's-law-type arrangement, which is shown to be an arrangement that minimizes interaction action.

According to Ovenden (1975), the present observed arrangement of the planets is not one for which interaction action is a minimum. The problem is the so-called missing planet in the asteroid belt at 2.8 AU. From the present arrangement of the major planets, Ovenden has estimated the presence of a planet with a mass of order $10^2 \, M_\oplus$ in the asteroid belt until approximately 10^7 years ago. At that distant time the planet, named Krypton, broke up in some as yet mysterious manner. Krypton would have had a composition like the Jovian planets rather than like the terrestrial planets. Thus, it is likely that its breakup would have ejected icy masses into space that look much like our conception of the nucleus of a comet.

Originally a rather violent end was envisioned for Krypton. However, it is widely felt that an event of such magnitude would have left its mark throughout the solar system. The current concept involves a more gradual breakup or disruption, in which case the fate of the disrupted material is probably known – 99.9% of it was captured by Jupiter. The addition of about $100 \, M_\oplus$ to Jupiter (30% of its present mass) may provide an explanation of Jupiter's outer satellites, which could not be captured by Jupiter without invoking important dissipative forces. These satellites might have been fragments of the disrupted planet that would not have been permanently captured by Jupiter except through the effect of a major increase in Jupiter's mass. Ovenden has also suggested the breakup of the missing planet as the origin of the asteroids.

The views put forth by Ovenden are the subject of substantial criticism and debate and should be viewed as highly speculative. Both the principle of least interaction action and Ovenden's conclusions will stand or fall on their own. Our interest here is in the extension of these ideas to comets by T. C. Van Flandern (1978). He suggests that the long-period comets are the high-energy residue of the breakup of Ovenden's missing planet.

Van Flandern's approach naturally explains the fact that observed periods of the long-period comets coming to the inner solar system for the first time are of order 10^7 years. It also explains the excess of retrograde over direct orbits among the long-period comets (Chapter 3). Apparently, most of the high-energy fragments escaped from the solar system. However, escape would be easier for direct comets because the orbital motion of the planet would be added to them and subtracted from the retrograde comets.

There are approximately 60 very-long-period comets with well-determined orbits. These are the so-called first-around comets, and they may be used to test the planetary model of cometary origin. To the extent that the original orbits have not been perturbed, they should intersect at the same point on the celestial sphere – their point of origin. The orbits of the 60 first-around comets during their most recent apparition can be checked. A significant clustering is found near 249° ecliptic longitude (and at the diametrically opposite point on the sphere). The principal perturbations on the orbits are the randomizing effects of stellar encounters and the tidal effects of the galactic field. The latter can be removed by numerical integration back in time to the previous perihelion passage, taken to be 6×10^6 years ago. The orbits cluster even more near ecliptic longitude 258°, and close to the ecliptic. The scatter in the orbits is entirely consistent with the effects of random stellar encounters. One should note that the clustering of intersections is a geometrical effect. There is no evidence to indicate that all comets were at the intersection points at the same time.

Van Flandern also points out that in the breakup of Krypton, more particles will have initial velocities nearly perpendicular to the radius

Figure 5.15. Perihelion directions of 60 very-long-period comets, in ecliptic coordinates centered at longitude 165°. (Courtesy of T. C. Van Flandern, U.S. Naval Observatory)

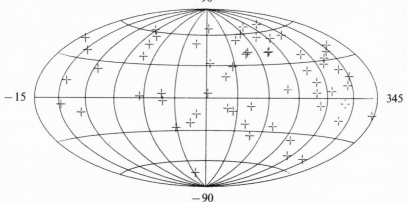

vector from the sun rather than nearly parallel to it, because those velocities nearly perpendicular can have an arbitrary component. Under these circumstances, for particles in nearly parabolic orbits the sun ejects approximately twice as many particles toward one hemisphere of the sky as toward the other hemisphere. The hemisphere centered on longitude 258° contains twice as many perihelia as the other hemisphere (see Figure 5.15).

Finally, the Oort cloud model makes a prediction that is not in agreement with the observations. If the comets in fact circle the sun at distances like 60,000 AU for a long time and are sent into the inner solar system by stellar perturbations, their angular momentum should be randomized by the stellar encounters. For parabolic orbits, the perihelion distance q and the angular momentum h are related by

$$q = h^2/2\mu \tag{5.15}$$

where $\mu = G(M + m)$. In this situation, there should be approximately the same number of comets on equal intervals of $h/(2\mu)^{1/2}$. For the interval between 0.7 and 1.0 (q between 0.49 and 1.00 AU) and the interval between 1.0 and 1.3 (q between 1.00 and 1.69 AU) this is not the case. Even though the probability of discovery is higher for the smaller q interval, it contains 9 comets versus 20 for the higher q interval. This result has a probability of less than 0.03 to occur by chance and is an additional argument against the traditional view. However, Delsemme argues that those comets that reach smaller q values vaporize more rapidly per passage than larger q value comets, and therefore decay and disappear more rapidly. Thus, he predicts fewer comets in the smaller q value range.

Van Flandern's evidence relates almost entirely to the orbital dynamics side of the origin of comets. Even if this aspect is correct, it would then be necessary to understand the breakup of the missing planet and the processes that produced comets from the energetic fragments. If comets are due to a specific event in the fairly recent past (on the scale of astronomical times), we may be witnessing a unique display in the evolution of planetary systems. The reader is advised to keep up with recent literature for further developments in this area.

6   *Comets and the solar system*

Relation to meteor streams

The historical connection between comets and meteor streams was established in the nineteenth century, when it became clear that the orbits of some meteor streams were essentially the same as the orbits of certain comets (see the historical account in Chapter 2 and Figure 6.1). Some major showers related to specific comets are listed in Table 6.1. The associations listed in Table 6.1 are considered to be relatively certain. The showers are named for the location of their radiants on the celestial sphere; the older practice of naming showers after the associated comet has been discontinued. The shower dates are approximate, and the intensity of the shower can vary greatly depending on whether the material is uniformly distributed along the orbit or in clumps. Bear in mind, in addition, that the meteor stream can be perturbed by Jupiter, for example.

There are many streams without an associated comet, but the current view assigns a parent comet in the past to all such streams. The so-called sporadic meteors appear to come from a background level of interplanetary material thought to have originated in streams that have been dispersed. The evidence indicates that the stream and nonstream particles are the same physically. The meteoroids of cometary origin probably account for about 99% of the material encountered by the earth in its orbital motion.

Thus, the study of meteors is, in effect, the study of the solid material in the cometary nucleus. Unfortunately, there is no firm evidence to indicate that a large piece of a cometary meteoroid has ever reached the earth's surface. Although at first this is disappointing, this fact contains information and upon reflection is not surprising. If the cometary nucleus is composed principally of water ices, the residue after the icy material sublimates should be a porous, fragile structure. This viewpoint is strongly confirmed by the densities of cometary meteoroids as derived from the physical theory of meteor luminosity. Results based on 324 meteors photographed at the field stations of the Harvard College Observatory give an average density for cometary meteors of 0.3 g cm^{-3}. About 15% of the sample had higher densities, in the range 1–2 g cm^{-3}; these densities resulted from meteoroids with orbits lying entirely within the orbit of Jupiter.

Although the absence of meteorites of cometary origin precludes a

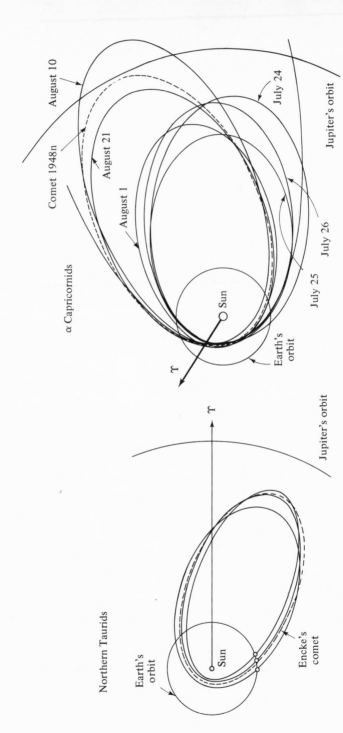

Figure 6.1. Orbits for meteors of two showers (several appearances) compared with the orbits of the associated comets. The agreement is excellent for the Northern Taurids and Encke's comet and for the August stream of the α Capricornids. The July stream of the α Capricornids may have another origin. (After J. C. Brandt and P. W. Hodge)

direct determination of their chemical composition, the study of meteor spectra presents such an opportunity. A sample meteor spectrum is shown in Figure 6.2. The results for elements identified have been summarized by Millman (1971:157) as follows: "Taking the lists for both comets and meteors we find the common elements we might expect to appear in the range of wavelengths accessible to astronomical observation, and under conditions of fairly low excitation potentials."

The identification of an element is trivial compared to the quantitative determination of relative chemical abundances. The problem is not amenable to straightforward theoretical interpretation because of the large number of poorly known mechanisms and because conditions are very far from equilibrium. Laboratory experiments under simulated meteor conditions have determined absolute luminous efficiencies for Na, Mg, Ca, and Fe. The laboratory calibration combined with the accurate photometry available for a few comets yields the relative abundances given in Table 6.2. The abundances are by weight and are normalized to 45. The conditions for good abundance determinations are satisfied for the Draconid meteoroids but not for the Perseid meteoroids. The composition of the cometary meteoroids shows good agreement with the relative abundances of meteorites for the elements listed.

The lighter elements expected from our picture of the cometary nucleus have also been observed in the spectra of meteors. The lines of atomic hydrogen, oxygen, and nitrogen have been observed. Of course, the results require a separation of the contribution of the atmospheric gases for O and N. Molecular bands of N_2, CN, C_2, and possibly CH have also been identified in meteor spectra. Although the luminous efficiency factors for the lighter atoms and molecules are not available,

Table 6.1. *Comets and meteor showers*

Comet	Shower	Duration
1861 I	Lyrids	April 20–23
1835 III (Halley)	η Aquarids	May 3–4
1835 III (Halley)	Orionids	October 18–26
1957c (Encke)	Daytime β Taurids	June 24–July 6
1957c (Encke)	Taurids	September 15–December 2
1862 III	Perseids	July 29–August 17
1948n	α Capricornids	August 1–21
1946 V (Giacobini–Zinner)	Draconids	October 9–10
1852 III (Biela)	Andromedids	November 2–22
1866 I (Tempel–Tuttle)	Leonids	November 14–20
1939 X (Tuttle)	Ursids	December 17–24

it seems highly probable that significant amounts of H, C, N, and O are present in cometary meteoroids and that the mass fractions are higher than in the meteorites.

Recently it has become possible to sample the composition of cometary meteoroids directly by instruments flown in the earth's atmosphere during a meteor shower. This was done by Goldberg and Aiken on July 3, 1972, for the β Taurid meteor shower that is associated with Comet Encke. The region of measurement was from 85 to 120 km altitude. Metallic ions exist at these levels, and the density enhancement during meteor showers apparently establishes their extraterrestrial origin. In

Figure 6.2. A bright Perseid meteor spectrum. The meteor moved downward in this presentation from an altitude of 114 km to 75 km at the end.
Lines of elements are labeled; note the 5577 Å line of oxygen at the top.
(Courtesy of Peter M. Millman, Herzberg Institute of Astrophysics)

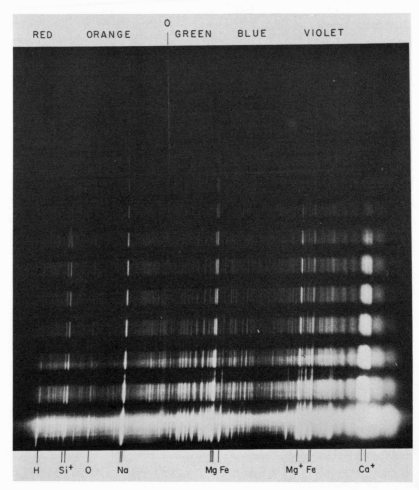

addition, the measured compositions are far closer to the chondrites than to the earth's crust. The ions measured originate during and following the ablation of the cometary meteoroids in our atmosphere.

The measurements were taken with an ion mass spectrometer and are shown in Table 6.3. The data are given for two heights in the atmosphere and for chondrites as a comparison; the abundances are normalized to Si^+, $Si = 100$. The relative abundances are reasonably close to chondritic abundances except for Sc^+. This is the first identification of species with a mass-to-charge ratio of 45 at these heights in the atmosphere, and hence the identification must be regarded with some suspicion. Nevertheless, no viable alternate source has been suggested, and we must consider the possibility that scandium is rather abundant in Comet Encke's solid material.

Comparison of this table with Table 6.2 for relative abundances derived from meteor spectra shows the strength of the in situ method. It is possible to measure elements far less abundant than those detectable

Table 6.2. *Relative abundances of elements in meteorites and cometary meteoroids*

	Na	Mg	Ca	Fe	Total
Meteorites	0.7	13.9	1.4	28.6	45
9 Draconid meteoroids	1.3 ± 0.1	13.2 ± 0.6	1.2 ± 0.1	29.3 ± 0.5	45
2 Perseid meteoroids	0.5	14.5	5.0	25.0	45

Source: After Millman (1971).

Table 6.3. *In situ meteor ion abundances as compared to chondrites*

Element	Ions		Neutrals: chondrites
	101 km	114 km	
Na	12.0	9.1	3.8
Mg	49.0	57.0	81.0
Si	100.0	100.0	100.0
K	0.81	0.38	0.51
Ca	3.6	1.2	7.9
Sc	1.0	0.35	0.005
Cr	3.4	1.8	1.7
Fe	149.0	158.0	141.0
Ni	4.2	4.7	7.6

Source: Data from Goldberg and Aikin (1973).

by optical means and, considering the difficulties of interpretation, it ultimately offers the hope of more accurate determinations. The abundances of the trace elements (such as Sc) could offer valuable clues to the origin of comets if a sufficiently large sample can be obtained.

Relation to the zodiacal light

A variety of hypotheses has been suggested for the origin of the dust responsible for the zodiacal light. These include: (1) dust from comets; (2) fragments from the collision of asteroids; (3) remnants of the solar nebula; and (4) dust in an interstellar cloud through which the solar system is now passing. The last two hypotheses are not viable at the present time, and the first hypothesis is preferred. If the zodiacal dust were supplied by collisions among asteroids, the concentration in the plane of the ecliptic would probably be too high. Hence, comets are the logical choice, but this conclusion is not without quantitative difficulties.

Whipple (1955) has long suggested that comets would supply the estimated 10^7 g sec^{-1} of solid material necessary to maintain the zodiacal cloud in a steady-state condition. A recent report by Giese and Grün (1976) shows that the light from the zodiacal cloud is dominated by particles larger than a few tens of microns in diameter. The zodiacal cloud is destroyed by the Poynting–Robertson effect, erosion, and collisions; the estimated lifetime for the entire cloud is $\sim 10^5$ years and the total mass is $\sim 2 \times 10^9$ g. On Whipple's model, the solid material was supplied by the short-period comets, principally P/Encke.

However, the Finson–Probstein model of dust emission from comets (Chapter 4) indicates that the dust production as seen in the type II tails is somewhat lower than assumed earlier. Hence, the short-period comets probably cannot supply the required solid material unless it is in the form of large pieces, such as boulders, which can escape detection by their reflected light. At present a population of boulders from comets is the best candidate for the source of the zodiacal light material.

We should not, however, overlook the possible past contribution from Comet Encke. It was surely much more massive in the past and has impressive meteor activity associated with it. Hence, as noted by Sekanina, Comet Encke may have been a major contributor to the zodiacal light in past epochs.

Comets as probes of the solar wind

The orbital characteristics of comets make them excellent candidates for natural probes of the solar wind. This fact was noted in an interplanetary context by Barnard in 1899. There are several possible ways that comets could be used to study the solar wind.

1. Brightness fluctuations of comets might be physically related to properties of the solar wind. This possibility has been reviewed by Miller (1976), and he has concluded that useful results were not obtainable, at least at present.
2. The acceleration of knots and kinks in plasma tails might be related to solar wind properties. Historically, Biermann (1951) used the accelerations of knots as an argument for the existence of the solar wind. In principle, the kinematic behavior of fine structure in the plasma tail could yield information about the solar wind, but unfortunately, we are in no position to utilize this information at present. For example, there is no general agreement even on the significance of the observations, namely, are the knots the result of bulk motion or of wave motion?
3. The orientation of the plasma tail appears to give reliable information on the solar wind velocity. A study of the orientations of the plasma tails of comets was published by Hoffmeister in 1943. He showed that the tails lagged (in the sense of the comet's orbital motion) behind the prolonged radius vector by an amount proportional to the comet's orbital velocity perpendicular to the radius vector, $v_\perp$. The angle between the plasma tail and the radius vector is called the *aberration angle*, ϵ, and is given by tan $\epsilon \approx v_\perp/w_r$, where w_r is the radial speed of the solar wind. The average value of the aberration angle is about 5°. These facts were used by Biermann in his papers, which, in essence, were the discovery of the solar wind. As of this writing, the orientations of plasma tails are apparently the only reliable, quantitative cometary indicator of solar wind conditions.

The earlier work on solar wind properties from comet tail orientations was based on the assumption of coplanarity, that is, that the tail was in the plane of the comet's orbit. This simplifying approximation converted a basically three-dimensional problem in space to a two-dimensional set of observables on the plane of the sky as recorded on the photographic plate or film. This simplified approach is not used in the modern work, except to derive a minimum value of the solar wind speed.

The shape (and orientation) of the plasma tail in this *wind sock* theory is determined by the local direction of momentum flow as seen by an observer riding on the comet. The basic equation is

$$\mathbf{T} = \mathbf{w} - \mathbf{V} \tag{6.1}$$

where $\mathbf{T}$ is the direction of the tail, $\mathbf{w}$ is the solar wind velocity, and $\mathbf{V}$ is the orbital velocity of the comet. On the wind sock theory, the tail is attached to the nuclear region by the magnetic field lines that thread the tail and are anchored in the cometary ionosphere.

Application of equation 6.1 to observations of the plasma tail's orientation at the head yields an average model of the solar wind velocity field. The basic observational quantity is the position angle (Θ) of the tail axis as measured on the plane of the sky (see Figure 6.3). In the astrometric approach, these are compared with calculated position angles (Θ_{cal}) generated for any assumed velocity field through equation 6.1 and the necessary geometrical transformations. The model is derived by utilizing standard least-squares techniques to minimize the quantity

$$\sum (\Theta - \Theta_{cal})_i^2 \tag{6.2}$$

In practice, the forms of variations in the solar wind components are chosen on the basis of theory and some available spacecraft observations. These are introduced parametrically as follows:

$$
\begin{aligned}
w_r &= \text{const} \\
w_\phi &= w_{\phi,o} \frac{(\cos b)^{2.315}}{r} \\
w_\theta &= w_m \sin b
\end{aligned}
\tag{6.3}
$$

Here w_r is the radial solar wind speed; w_ϕ is the azimuthal speed with $w_{\phi,o}$ the value in the plane of the solar equator at 1 AU; b is the solar latitude and r is the heliocentric distance; w_θ is the polar speed, with w_m the maximum value at $b = \pm 45°$. These variations are not sacred and, indeed, any model that can be represented by a few parameters can be tested by the astrometric technique.

The best current models for the global, long-term velocity field are given in Table 6.4. The dispersions given correspond to an isotropic, peculiar velocity in the range 30 to 50 km sec^{-1}. Also, a minimum solar wind speed of 225 ± 50 km sec^{-1} is derived from comet tail data. The numbers listed here are the only global, long-term values available for the solar wind velocity field. All the quantities listed are in agreement or consistent with the values obtained from spacecraft.

A major source of uncertainty is a possible variation of radial solar wind speed with latitude. Some radio scintillation studies and current ideas on solar wind flow from coronal holes require much faster solar wind speeds over the solar poles corresponding to a gradient in latitude of approximately 2 to 3 km sec^{-1}-deg^{-1}. If such a gradient were a long-term, global property of the solar wind, it should be evident in the data.

Table 6.4. *Solar wind models derived from comet tails*

	Main sample	Augmented sample
w_r (km sec^{-1})	402 ± 12	400 ± 11
$w_{\phi,o}$ (km sec^{-1})	7.0 ± 1.8	6.7 ± 1.7
w_m (km sec^{-1})[a]	+2.6 ± 1.2	+2.3 ± 1.1
RMS dispersion	3°749	3°736
No. of observations	678	809

[a] The positive values mean a flow diverging from the plane of the solar equator.

Figure 6.3. Comet West on March 9, 1976, showing the angles used on the astrometric approach. (Joint Observatory for Cometary Research, NASA–Goddard Space Flight Center and New Mexico Institute of Mining and Technology)

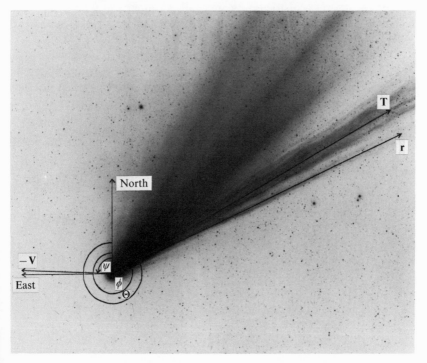

A solution with a latitudinal variation of the form

$$w_r = w_o + \frac{dw_r}{d|b|} \, |b| \tag{6.4}$$

has been tested. The gradient was found to be $dw_r/d|b| = -0.9 \pm 0.7$ km sec^{-1}-deg^{-1}, and, in addition, the root mean square (RMS) dispersion did not decrease. Statistically, this result means that in the sample tested the additional parameter has no significance.

Note that there is not necessarily a conflict between the comet tail results and the other evidence cited. There has been no opportunity to compare the results from cometary studies and radio scintillation studies for the same time period. Thus, the cometary results could be correct as an average over several solar cycles ($\approx$75 years) and the radio scintillation results correct for shorter periods of time. The results should converge to the same long-term values as scintillation data accumulate and as spacecraft explore the higher-latitude regions of interplanetary space, for example, by NASA's proposed International Solar Polar Mission.

Another reason for confidence in the cometary results comes from recent results on a distorted plasma tail. Comparison with solar wind speeds measured by spacecraft indicates that plasma tails respond to changes in speed with considerable fidelity.

Finally, recent morphological studies of plasma tails provide evidence that comets are useful as probes of the interplanetary magnetic field in addition to the velocity field data discussed above. A characteristic appearance of the plasma tail may allow the mapping of the magnetic sector boundaries in the solar wind. These ideas are discussed in Chapter 7.

Part III

The future perspective

In the past few years, cometary workers have had the good fortune to view
several bright comets. Examples are comets Kohoutek, Kobayashi–
Berger–Milon, and West. Bright comets not only produce public interest
but often inspire a higher level of activity in cometary research. In Part
III we review results from recent comets and developments that have taken
place concurrently. These developments have not yet withstood the test of time
and, generally, are less certain than those presented in Part II.

One of the exciting discoveries of the space age was the hydrogen cloud
surrounding comets. That discovery required an instrument on board an
earth-orbiting spacecraft. This result is one of many that have come from
studies of recent comets both from the ground and from space, with
ground-based observations still providing the majority of the information.

In the future, the relative contributions of ground-based and space-based
studies of comets may be reversed. However, to achieve their great potential,
spacecraft will have to leave the confines of earth's orbit and venture into deep
space to make in situ observations. Imagine a spacecraft flying in formation
with a comet, near the nucleus, watching the events that occur as the comet
moves in toward the sun and then out through the solar system, analyzing
samples of plasma as it goes. The answers such a mission would provide would
revolutionize our understanding of cometary phenomena. Is such a mission
feasible? Yes, we believe so. In this part, we describe in some detail the current
plans to fly by Halley's comet in 1985 and the possibilities for space missions to
comets.

Comet West in visible light (left) and in Lyman α (right) on the same scale. Both photographs were obtained from a rocket on March 5.5, 1976. (Visible photograph, courtesy of P. D. Feldman, Johns Hopkins University; Lyman α photograph, courtesy of C. B. Opal and G. B. Carruthers, Naval Research Laboratory)

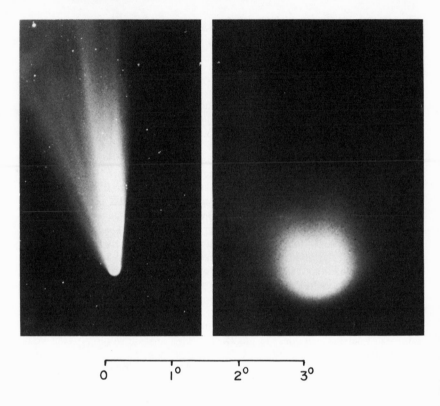

0   1°   2°   3°

7 Recent comets and some current developments

Kohoutek

The field of cometary physics has often been greatly affected by the passage of a particular comet. The most recent instance was the apparition of Comet Kohoutek (1973 XII). Comet Kohoutek was discovered well in advance of perihelion (Figure 7.1) and an extensive, coordinated observing program, including observations from above the atmosphere by *Skylab* astronauts, was carried out. Extensive radio observations, spectroscopy, and direct photography have vastly increased our fund of cometary knowledge.

Spectroscopic results

Radio observations in the millimeter wavelength range have led to the identification of two separate transitions of CH_3CN in Comet Kohoutek, as shown in Figure 7.2. The differences in intensities indicate that the populations in the excited states are probably not given by a Boltzmann distribution. The same wavelength range also yielded an identification of HCN, as shown in Figure 7.3. Finally, although water was not detected in Comet Kohoutek, it may have been detected in Comet Bradfield (1974b). These molecules are stable enough to survive for long periods of time in the nucleus, and all may be bona fide parent molecules of observed radicals. The molecules CH_3CN and HCN are probably minor constituents, but we have every reason to believe that H_2O is a major constituent of the nucleus. These results constitute the first direct detection of parent molecules. The spectra of HCN taken when Comet Kohoutek was close to the sun showed Doppler shifts that have been interpreted as due to jets with speeds as much as 3 km sec^{-1}.

In the centimeter wavelength range the radicals CH and OH, which earlier had been identified from optical spectra, were also found in Comet Kohoutek. The OH observations were particularly interesting because they were observed in absorption before perihelion and in emission after perihelion, as shown in Figure 7.4. These OH observations are the only observations of absorption lines in a comet.

Rocket-borne spectrographs have recorded the resonance lines of C, H, and O in the vacuum ultraviolet spectrum of Comet Kohoutek. This was the first detection of atomic carbon (Figure 7.5).

A substantial list of negative results (among the more interesting being CO) has also been obtained, as well as the detection of several unidentified lines.

Production rates for many species in Comet Kohoutek were determined. These rates, which have been collected by Delsemme (1976, 1977), have been referred to 1 AU using the inverse square law to produce the values given in Table 7.1. Among the interesting results in this table is the implication that the abundances of CN and C_2 with respect to water seem to be a factor of 10 or more higher for Comet Kohoutek than for an obviously old comet such as P/Encke. Delsemme has noted that this appears to be the first quantitative evidence of a large difference in the production rates between an old and a new comet.

Finally, optical spectra of Comet Kohoutek have led to the identification of many lines in the bands of the H_2O^+ molecule (see Figure 7.6). All major photodissociation and photoionization products of water, as well as the parent molecule itself, have been reported either in Comet Kohoutek or in Comet Bradfield.

Cometary atmosphere

Extensive observations of Comet Kohoutek's coma have given a new body of data for scale lengths, production rates, and so on. Analysis of these brightness profiles yields results, all of which are consistent with water as the major nuclear constituent.

Figure 7.1. The discovery photograph of Comet Kohoutek (1973f) taken on March 7.877 UT, 1973. (Courtesy of L. Kohoutek, Hamburg Observatory)

Figure 7.2. Discovery observations of CH₃CN (methyl cyanide) in Comet Kohoutek. (*a*) December 1, 1973; (*b*) December 5, 1973. The bottom spectrum is the average of (*a*) and (*b*). (Courtesy of B. L. Ulich and E. K. Conklin)

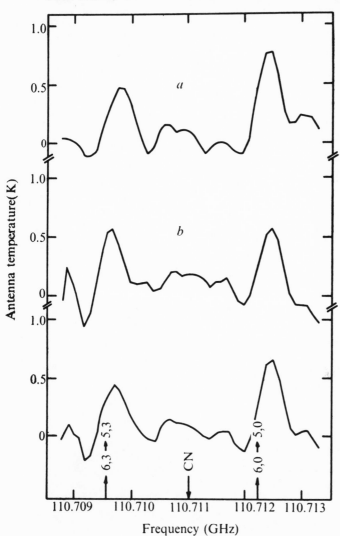

Figure 7.3. Discovery observations of HCN (hydrogen cyanide) in Comet Kohoutek. Each horizontal bar connects triplet components belonging to the same Doppler-shifted jet. Bars with arrows indicate the "quiescent" emission triplet, whereas dashed bars connect triplets not significantly above noise level. The vertical bars indicate the noise level with the dome closed (dotted bars), and the solid bars are the dotted bars multiplied by 1.6. The percentage of time for which observations were carried out with the dome open are displayed to the right of each spectrum. (Courtesy of W. F. Huebner, L. E. Snyder, and D. Buhl)

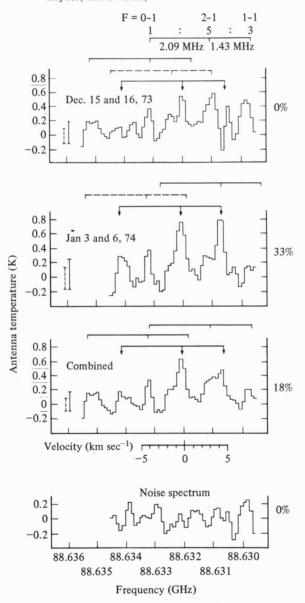

Figure 7.4. (*a*) The peak antenna temperature in the 1667-MHz line of OH in Comet Kohoutek as a function of time. The line is in absorption in early December 1973 and reappears in emission around mid-January 1974. (*b*) The predicted variation as a function of time on the basis of ultraviolet pumping. The variation is given in terms of the populations of the upper level (n_u) and the lower level (n_l), and the quantity $(n_u - n_l)/(n_u + n_l)$ is plotted. (Courtesy of F. Biraud, G. Bourgois, J. Crovisier, R. Fillit, E. Gérard and I. Kazès, Observatoire de Meudon)

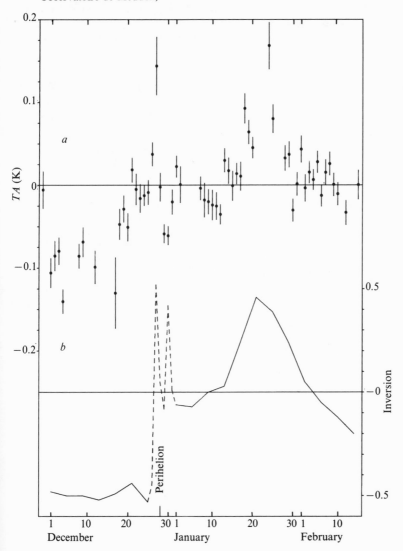

The ionized part of the atmosphere, or cometary ionosphere, is receiving increased attention. The need for rapid processes, called by Wurm *intrinsic cometary processes,* to form the molecular ions, notably CO^+ and N_2^+ quite near the nucleus, has long been recognized. One possibility is fast ion-molecule chain reactions in the inner coma; presumably, these would be initiated by photoionization. These ideas have been pioneered by Cherednichenko (1974) and elucidated recently by Oppenheimer (1975). These fast reactions would be confined to the inner 10^4 km of the coma, where collisions are important, and hence, if operative, would produce ions in a restricted volume around the nucleus. However, Oppenheimer emphasizes the possible complexity of the situation by listing some 100 reactions (half of these involving ions) that might be important in the inner coma and by declining to predict a steady-state model. Despite the uncertainties, several workers agree in assigning an important role to the H_3O^+ ion. Observing the spectrum of

Figure 7.5. Ultraviolet spectrum of Comet Kohoutek obtained on January 5.1 UT, 1974. Besides the atomic carbon (C I) lines at 1561 Å and 1657 Å, the spectrum also shows the off-scale features of atomic hydrogen (Lyman α) at 1216 Å and atomic oxygen at 1304 Å and 1356 Å. The solid curve is the instrumental response to a source of uniform intensity. (Courtesy of P. D. Feldman, P. Z. Takacs, W. G. Fastie, and B. Donn)

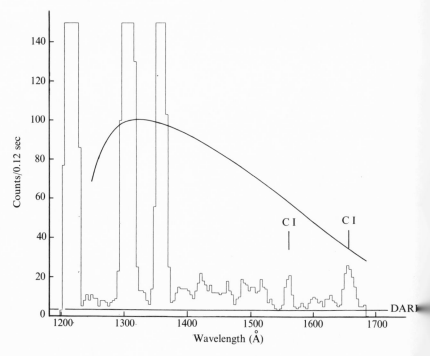

this ion in comets will require the laboratory determination of its spectrum, which is presently unknown.

The second possibility for production of the ions near the nucleus is the closing of currents generated in the tail through the inner coma. Direct photographs of Comet Kohoutek on January 13, 1974, showed a helical structure, one explanation of which involves large currents flowing along the ion tail. Apparently building on this suggestion, Ip and Mendis (1976) have proposed that large currents in the tail could close through the inner coma and provide an efficient ionization source. The idea is generally analogous to the situation in the geomagnetic tail; see Figure 7.7. A desirable feature of this mechanism is that the solar wind conditions producing the currents may not be a simple function of heliocentric distance, thus allowing an explanation of the prodigious generation of CO^+ by Comet Humason when located some 5 AU from the sun. Obviously, the two suggested mechanisms need not be exclusive.

Dust coma

Infrared observations of comets Kohoutek and Bennett showed the emission features (superimposed on the thermal continuum) near 10 and 18 μm commonly ascribed to small silicate grains (Figure 7.8). Note, however, that this interpretation has been questioned by Mendis and Wickramasinghe (1975). The thermal continuum is well represented by a black body curve in the wavelength range 2 to 20

Table 7.1. *Production rates in Comet Kohoutek*

Species	Rate (10^{27} particles sec^{-1})
H (from Lyman α)	25–40
H (from Hα)	4[a]
OH (from 3090 Å)	13–15
OH (from 18 cm)	4[b] or 500[b]
O I (1304 Å)	26
[O I] ^1D	11
C I (1657 Å)	6
C_2	2
CN	0.5
H_2O^+	1[a]
CH (9 cm)	300[b]
CH_3CN	≥ 1
HCN	≥ 2

[a] Uncertain result.
[b] *Very* uncertain result.
Source: Delsemme (1976, 1977).

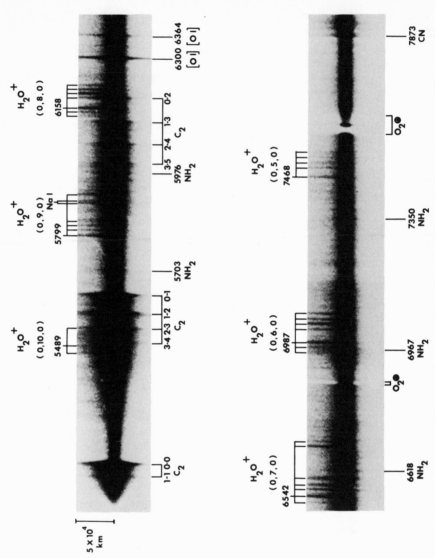

Figure 7.6. Optical spectrum of Comet Kohoutek showing H_2O^+ lines. Wavelengths in Angstroms. (Courtesy of S. Wyckoff and P. A. Wehinger)

Figure 7.7. Diagram of a possible cometary current system, with a cross-tail current and a solenoidal current. A dynamo is driven in the earth's magnetosphere due to the solar-wind plasma flow interacting with the magnetic field. There are two current systems in the magnetosphere: a system in the magnetotail that consists of a cross-tail current and of a solenoidlike current, and a system that discharges through the earth's polar upper atmosphere producing the aurora. The cometary cross-tail current, shown here, may be partially disrupted by instabilities, causing it to flow along field lines that are out of the ray-closing symmetry plane and near the neutral sheet (bold arrows) and to discharge through the coma, heating it. The vertical scale is exaggerated for clarity.

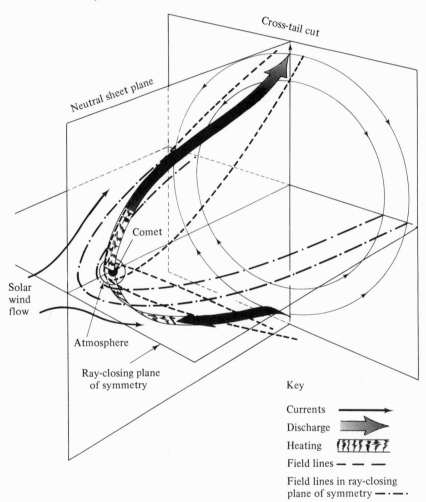

μm, and the temperature varies from 350 to 700 K depending on heliocentric distance. Considerable evidence has been accumulated that indicates variations in the dust composition from comet to comet and for a particular comet with time. The usual dust coma appears to have been detected in millimeter wavelengths by Bruston et al. (1974).

Comet Kohoutek was detected in continuum radiation at radio wavelengths. This appears to be the first such detection of any comet. Hobbs et al. (1975) detected the comet at 3.71 cm using interferometric techniques. The detection is firm, and the probable origin is in an icy grain halo composed of dirty grains. Note that this detection refers to 1 day only, and an unusual state of the comet cannot be excluded.

Dust tail and antitail

A prominent antitail was observed for Comet Kohoutek both from *Skylab* (Figures 4.9 and 7.9) and from the ground (Figure 7.10). According to Sekanina (1977), the only requirements for an antitail or sunward-appearing dust tail are (1) a favorable geometry and (2) the presence of millimeter-sized particles. Because antitails seem to be a property of nearly parabolic comets and not short-period comets, it appears that young comets alone have sufficient volatiles to drag the

Figure 7.8. Infrared observations of comets Kohoutek and Bennett at the same heliocentric distance showing their black body temperatures and silicate features (10 to 18 μm). See also Figure 7.23. (E. P. Ney, 1974, *Icarus* *23:*551–60)

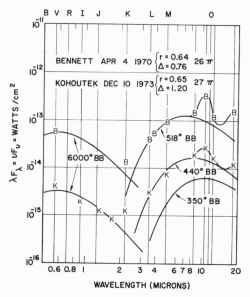

millimeter-sized particles away from the nucleus at the heliocentric distances at which comets are usually observed.

Application of the Finson–Probstein coma model (see Chapter 4) or equivalent models by several authors confirms the presence of millimeter-sized particles. In addition, there was evidence for vaporizing dust particles. The regular dust tail had a normal appearance, as shown in Figure 7.10.

The ion tail

Photographs of Comet Kohoutek in January 1974 showed spectacular tail activity; examples are shown in Figures 7.11 and 7.12. The observation of moving structures in the ion tail has reopened the debate

Figure 7.9. Comet Kohoutek with its antitail as sketched from *Skylab* on December 29, 1973. (Courtesy of E. G. Gibson, NASA–Johnson Space Center)

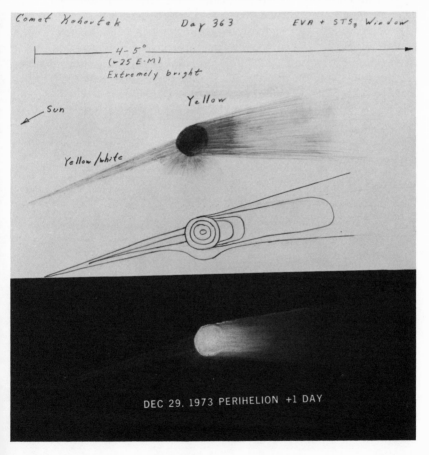

concerning their interpretation. Both the helical structure observed on January 13 (Figure 7.11) and the Swan cloud observed on January 11 (Figure 7.12) are moving away from the head at ≈ 200 km sec^{-1}. They could be waves triggered as a result of the kink instability and traveling at the Alfvén speed of the cometary plasma. This would imply magnetic fields well down the tail (0.1 AU from the head) of 100 γ or more.

The alternate interpretation ascribes these moving features to bulk motions originally triggered by a Kelvin–Helmholtz instability at the solar wind–cometary plasma interface. This picture assumes that the magnetic field in the tail is approximately the same as the magnetic field in the local solar wind, or approximately 20 to 25 γ. The magnitude of the tail field is the key difference in these two interpretations.

Some support for the wave picture comes from the apparent observation of radio scintillations attributable to them by Ananthrakrishnan et al. (1975). The observed irregularities are propagating *toward* the nucleus at speeds in the range 66 to 94 km sec^{-1}. Ip and Mendis (1975) have interpreted this as a hydromagnetic wave propagating toward the nucleus with the Alfvén speed. They also provide an explanation for the lack of observation via radio scintillation of the Alfven waves propagating away from the nucleus. The geometrical circumstances indicate that the period of the fluctuations would be roughly 1 sec or less, and this period is the same as the rapid background interplanetary scintilla-

Figure 7.10. Photograph of Comet Kohoutek on January 14, 1974. Although the antitail is not as prominent as in Figure 7.9, it is still clearly visible (arrow). (Joint Observatory for Cometary Research, NASA–Goddard Space Flight Center and New Mexico Institute of Mining and Technology)

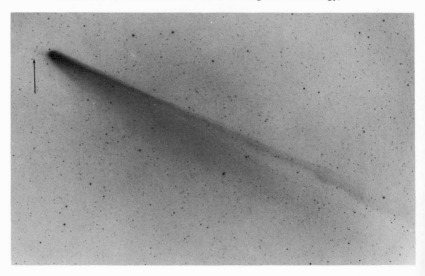

Figure 7.11. Comet Kohoutek on January 13, 1974, showing a helical structure. (Joint Observatory for Cometary Research, NASA–Goddard Space Flight Center and New Mexico Institute of Mining and Technology)

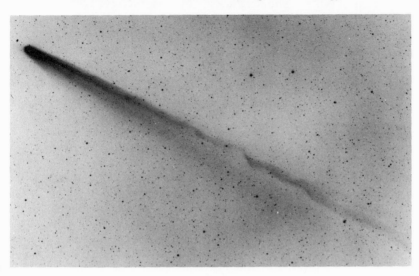

Figure 7.12. Comet Kohoutek on January 11, 1974, showing the Swan cloud some 0.1 AU from the head. (Joint Observatory for Cometary Research, NASA–Goddard Space Flight Center and New Mexico Institute of Mining and Technology)

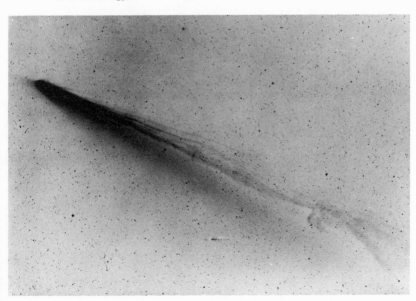

tions that mask the cometary effects. The fluctuation period for the irregularities propagating toward the nucleus is ≈ 10 sec, and hence they are observable. The length scale of the structures responsible for the scintillations is ≈ 140 km, a value much smaller than the resolution of optical telescopes. Therefore, at present these radio observations provide the only method of observing this level of fine structure in the plasma tails.

Ip and Mendis have also estimated the magnetic field in comet tails from the picture of folding field lines and conservation of magnetic flux. The equation is

$$B_t \approx \frac{B_0 w \, \tau_f}{h_t} \tag{7.1}$$

where B_t is the field in the tail, B_0 is the field in the solar wind, w is the solar wind speed, τ_f is the characteristic time for the folding of the tail rays (≈ 14.5 hr, according to Wurm and Mammano), and h_t is the half-width of the main tail. For conditions near 1 AU, $B_0 \approx 5\gamma$, $w \approx 300$ km sec^{-1}, and adopting $h_t \approx 5 \times 10^4$ km, we find $B_t \approx 1500\gamma$.

An alternate estimate is obtained simply by setting

$$B_e w \approx B_t v_t \tag{7.2}$$

where v_t is the folding velocity of the tail streamers. The streamers' folding velocity is in the range 50–100 km sec^{-1} away from the tail axis and decreases to ≈ 1 km sec^{-1} as the streamer nears the main tail axis. If we choose $v_t \approx 1$ km sec^{-1}, a B_t of 1500γ is obtained.

These calculations are quite simple, as Ip and Mendis have noted, and more detailed calculations are needed. However, the calculations suggest that magnetic fields in comet tails of 100γ or more are not unreasonable at this time. Nevertheless, Ershkovich (1978) has presented sound arguments for $B_t \approx B_0$.

The folding of the tail streamers to the tail axis can be interpreted as the folding of magnetic flux tubes into the tail. Ip and Mendis have suggested that this situation produces a cross-tail current that passes through a neutral sheet separating the two plasma regions of opposite magnetization. The total cross-tail current separating the oppositely magnetized regions is given by

$$I \approx \frac{c B_t l_t}{2\pi} \tag{7.3}$$

where B_t is the magnetic field in the tail and l_t is the total length of the neutral sheet along the tail. For $B_t \approx 10^3 \gamma$ and $l_t \approx 10^6$ km, we find $I \approx 10^9$ A. Ip and Mendis have developed this model further and find that if 20% of the cross-tail current or 2×10^8 A closes through the inner coma (within 10^3 km from the nucleus), it could produce ionization with a time scale $\approx 5 \times 10^3$ sec. Thus, the closing through the inner coma of large currents generated in the plasma tails could provide the long-missing source of ionization for the tail plasma.

Although we have presented arguments for interpreting moving structures in tails as waves and for large magnetic fields in the plasma tails, the situation is not settled. A direct measurement of the plasma motions in the tail, either by an in situ probe or more probably in the near future by the Doppler effect, is probably necessary for a measure of confidence.

The response of the plasma tail to changes in solar wind speed has been the object of a recent study of Comet Kohoutek. Wide-field photographs of the comet on January 20, 1974 (Figure 7.13), show a large-scale disturbance, a "bend," in the outer regions of the tail. The position angles of the various segments of the tail were measured and were combined with spacecraft (*IMP 8*) measurements of travel speeds to generate a signature of position angle versus time.

All three components of solar wind speed were measured continuously by *IMP 8*, and Comet Kohoutek's position near the plane of the

Figure 7.13. Comet Kohoutek on January 20, 1974, showing a large "bend" in the plasma tail. (Joint Observatory for Cometary Research, NASA–Goddard Space Flight Center and New Mexico Institute of Mining and Technology)

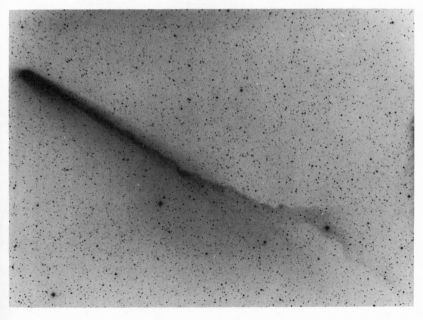

ecliptic was favorable for comparison with the spacecraft data. It is a straightforward task to generate the position angles that the hourly averaged solar wind speeds would produce at the comet. The only significant uncertainty is the time delay between the earth and the comet. The time delay is composed of terms due to the differences in solar longitude and heliocentric distance, and it is computed assuming that solar wind conditions remained constant. The standard formula gives a value of 4.4 days for the time delay and leads to the comparison between the observed and computed position angles as shown in Figure 7.14. The two sets of data can be made nearly coincident with a shift of 0.5 to 0.75 days in the time delay and a change of 6 km sec^{-1} in the polar component of solar wind speed. These adjustments are well within the uncertainties and, thus, a specific solar wind event has been linked to a specific structure in a plasma tail.

The solar wind event is a large, rapid change in the polar component of the velocity. The geometrical circumstances on January 20, 1974, were such that the shape of the plasma tail as viewed from earth was sensitive to the polar component. The change was associated with the

Figure 7.14. Observed position angles (solid lines) versus the computed position angles (dots) from *IMP-8* data corotated to the comet. See text for discussion. (Courtesy of M. B. Niedner, E. D. Rothe, and J. C. Brandt)

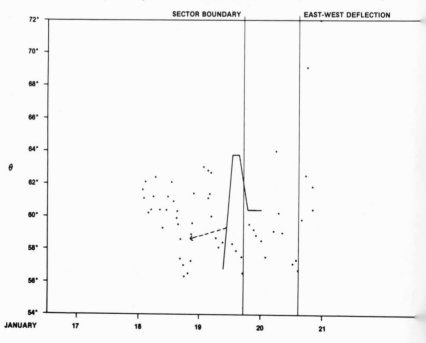

compression region of the high-speed stream that produced a geomagnetic storm on January 24–27, 1974. The solar origin of this stream may have been the large equatorward extension of the south polar coronal hole, which underwent central meridian passage on January 22. Clearly, this event illustrates the utility of comets as probes of the solar wind.

Finally, there is a good possibility that comets may also serve as probes of the interplanetary magnetic field. So-called disconnection events in which a comet apparently discards its tail have been described in the cometary literature for decades. The earliest known events were described by Barnard in the 1890s. The disconnection events could be produced when a comet crosses an interplanetary sector boundary (where the magnetic field changes direction from predominantly toward the sun to away from it, or vice versa). This possibility became apparent in the study of the January 20, 1974, event in Comet Kohoutek described above. The good correspondence between the calculated and observed position angles in the plasma tail (shown in Figure 7.14) provided confidence that the solar wind conditions at the comet were known around the January 20 date. Approximately 8 hours after the January 20 event, a disconnection occurred and the discarded tail was recorded on plates taken on January 21. The disconnection occurred shortly after the comet had crossed a sector boundary.

Although the January 20–21, 1974, disconnection event in Comet Kohoutek was relatively unspectacular, a search of the literature has turned up over 40 other events. Four examples are shown in Figure 7.15. Almost all these events are associated with sector boundary crossings as determined in one of two ways: (1) direct determinations are available for the modern events; (2) indirect determinations were made for the older events using the correlation between certain geomagnetic indices and solar wind speed. Approximately 15 years of in situ measurements of the solar wind have established the position of the sector boundary with respect to the characteristic variation of solar wind speed within the sector. This method can be used back to 1868.

A qualitative explanation of the disconnection phenomenon within the framework of Alfvén's theory of tail formation is possible and is illustrated in Figure 7.16. The basic new feature is the reversal of the interplanetary magnetic field (the sector boundary) encountering the comet. Magnetic reconnection cuts field lines anchored in the vicinity of the nucleus and eventually strips away the magnetic flux from the old sector to produce a completely disconnected plasma tail. At the same time, a new tail is formed with the flux from the new sector. As the process continues, the tail lengthens to form a normal plasma tail.

If this explanation of the disconnection events stands the test of time,

Figure 7.15. Four examples of disconnection events. (*a*) Comet Borrelly, July 24, 1903 (Yerkes Observatory photograph); (*b*) Halley's comet, June 6, 1910 (Yerkes Observatory photograph). A time sequence of this same event is given in Figure 2.3.

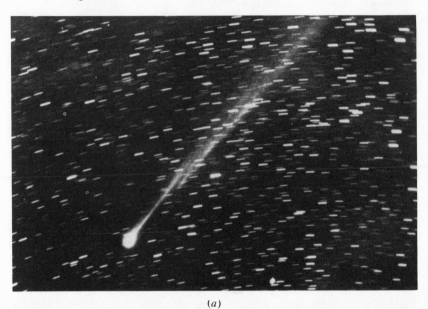

(*a*)

(*b*)

Figure 7.15 (*cont.*)
(*c*) Halley's comet, May 13, 1910 (Lowell Observatory photograph); (*d*) Comet
Bennett, April 4, 1970 (photograph taken by K. Lübeck, Hamburg
Observatory). (Courtesy of M. B. Niedner and J. C. Brandt)

(*c*)

(*d*)

Figure 7.16. The sector boundary–reconnection model of disconnection events; see text for discussion. (Courtesy of M. B. Niedner and J. C. Brandt)

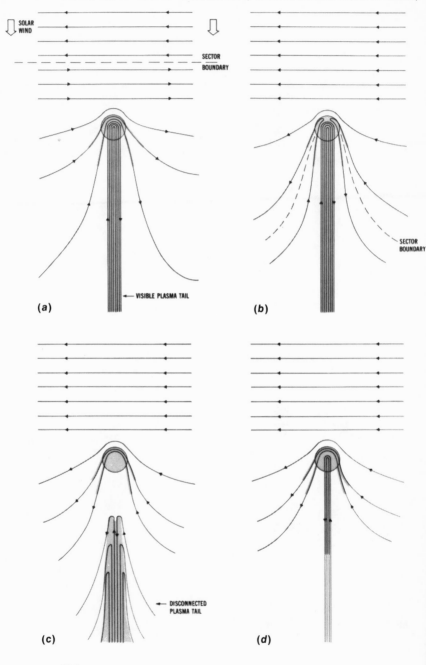

comets will indeed be able to map the interplanetary sector boundaries. These events may also be an important ionization mechanism for production of the ions found in the tail.

Kobayashi–Berger–Milon

Comet Kobayashi–Berger–Milon (1975h) was discovered as an eighth-magnitude object in early July 1975 by the successful young Japanese comet searcher Toru Kobayashi. Confirming observations were made by D. Berger and D. Milon during a siege of poor observing weather in Japan. The comet was brightening rapidly and reached naked-eye brightness less than 2 weeks after discovery.

The comet passed within 0.26 AU of the earth about 6 weeks before perihelion. Between near-earth passage and perihelion, the increase in the absolute brightness of the comet almost exactly compensated for the effect of increasing distance from the earth, and the comet maintained a nearly constant apparent brightness for almost a month. As a result, the comet was very well observed.

Spectrograms of the coma in July revealed the presence of bands due

Figure 7.17. Comet Kobayashi–Berger–Milon on August 2, 1975. (Joint Observatory for Cometary Research, NASA–Goddard Space Flight Center and New Mexico Institute of Mining and Technology)

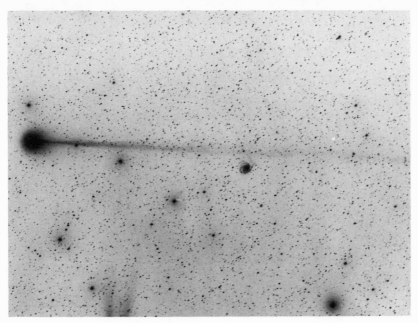

to CN, C_3, NH_2, NH, and OH. In addition, the forbidden line (O I) $\lambda 6300$ was present in the spectrum. Observations by the Copernicus satellite revealed an extensive Lyman α halo.

In early August the only detectable constituent in the tail appeared to be H_2O^+. However, the comet later developed a good ion tail (Figure 7.17). After perihelion the comet moved to high latitudes, and good time coverage of activity was obtained. The observations obtained on July 31, 1975 (Figure 4.13), provide an excellent illustration of the turning of tail rays and their lengthening as they approach the tail axis. These observations were made near in time to a tail disconnection event.

West

Stunning views of Comet West (1975n) in the morning sky were observed in early March 1976; see Figure 7.18 (cover photograph) and Figure 7.19. On several occasions the comet covered a roughly triangular area some $25° \times 25° \times 15°$ in the sky. It was easily visible to the

Figure 7.18. Wide-angle photograph of Comet West on March 4, 1976. (Courtesy of A. Stober, NASA–Goddard Space Flight Center)

naked eye even near large cities. The dust tail provided beautiful examples of the so-called synchronic band structure as shown in Figures 7.18 and 7.19.

This comet also showed unusual activity in the nuclear region. The comet apparently split three times and after March 5, 1976, consisted of four distinct nuclear components (Figure 7.20). One of these components had faded and disappeared from view by the end of March (Figure 7.20). The separations on the photographs are generally in the range 10,000 to 20,000 km, and the relative velocities increased with time for the interval shown in the range 10 km sec^{-1} to 40 km sec^{-1}.

The nuclear splitting of Comet West may help explain radio observa-

Figure 7.19. Mosaic of Comet West photographs taken on March 9, 1976. Note the development of the ion tail compared to the situation a few days earlier, shown in Figure 7.18. (Joint Observatory for Cometary Research, NASA–Goddard Space Flight Center and New Mexico Institute of Mining and Technology)

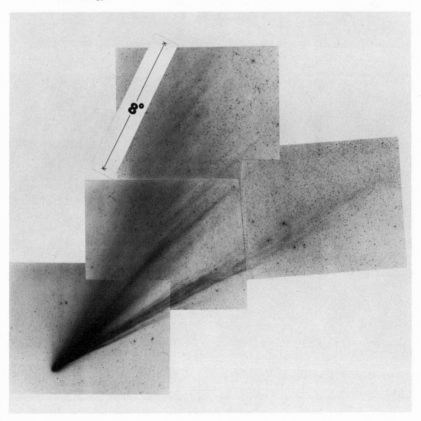

tions of the comet at 3.71 cm. This radiation was detected from Comet Kohoutek and interpreted as thermal radiation from the icy grain halo. Comet West was clearly detected on March 5, 1976, but not on March 4, 1976. The icy grain halo could have been enhanced on March 5 because of the nuclear splitting. If this interpretation is correct, the average icy grain halo has a much smaller mass and surface area than required to explain the detections at 3.71 cm. Thus, there may be no long-term problem concerning the total rate of vaporization.

Figure 7.20. Nuclear splitting in Comet West. (*a*) Views taken (left to right) on March 8, 12, 14, 18, and 24, 1976, in yellow-green light. On March 18, the projected diameter of the cluster was about 10,000 km. (*b*) Views taken (top row, left to right) on March 31, April 1 and 2, 1976, and (bottom row, left to right) April 3, 7, and 8, 1976, in red light. On April 7, the cluster was about 20,000 km long. (New Mexico State University Observatory)

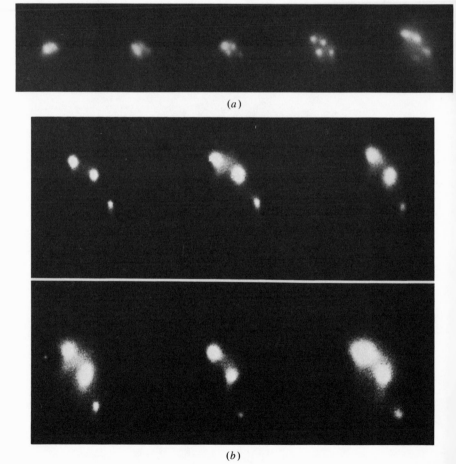

(*a*)

(*b*)

Figure 7.21. Ultraviolet spectra of Comet West (1975n) obtained on March 5, 1976. (*a*) Short-wavelength region. Note the improvement over the spectrum given in Figure 7.5. (*b*) Long-wavelength region. The solid curve is the instrumental response to a source of uniform intensity. (Courtesy of P. D. Feldman and W. H. Brune, Johns Hopkins University)

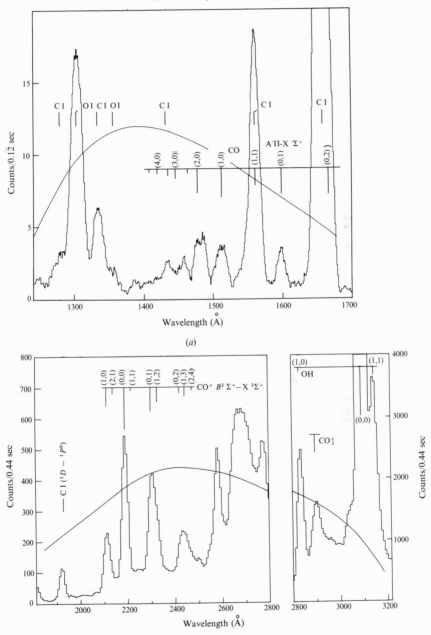

(*a*)

(*b*)

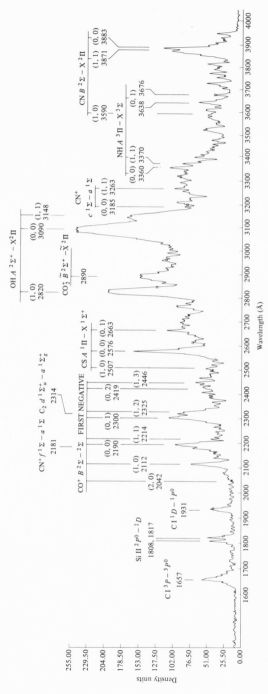

Figure 7.22. Density tracing of spectrometer observations of Comet West on March 10, 1976. (Courtesy of A. M. Smith, R. C. Bohlin, and T. P. Stecher, NASA–Goddard Space Flight Center)

Finally, there were several spectroscopic experiments carried out in the far ultraviolet by rocket-borne spectrographs and spectrometers. Carbon monoxide emission was positively detected for the first time. Besides CO and the species routinely detected, these observations detected C^+, CS, S^+, and SH. Sample spectra are shown in Figures 7.21 and 7.22.

Observations of the Lyman α halo (figure, first page of Part III) and infrared dust emission (Figure 7.23) of Comet West were also carried out. The dust emission showed increases in brightness that seem to coincide with the times of fragmentation.

Figure 7.23. Energy distribution of Comet West between February 2.8 and April 5.6, 1976. Scattered sunlight is the source of the short-wavelength radiation, whereas thermal radiation is the source of the long-wavelength radiation. The silicate feature is shown at 10 μm. Scattering angles are given in parentheses below the dates. Compare with Figure 7.8. (Courtesy of E. P. Ney and K. M. Merrill)

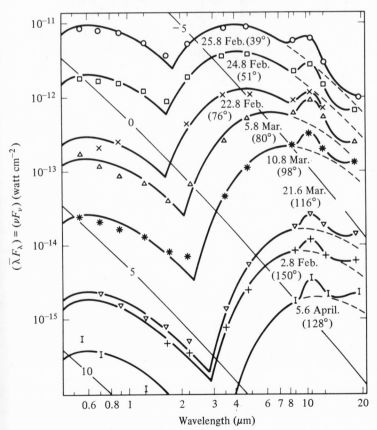

Bradfield

Comet Bradfield (1979l) displayed interesting changes in its plasma tail, as shown in the Frontispiece. Figure 7.24 shows a recent spectrum of Comet Bradfield. Note that the CO^+ molecular ion is not present in the spectrum.

Conclusion

These bright comets are a small sample of recent comets whose observations have added to our understanding. There is no

Figure 7.24. Spectrum of Comet Bradfield (1979l) obtained on February 5.1, 1980. This spectrum shows many of the principal cometary emissions in this wavelength range, but CO^+ is absent. A, nucleus; B, 43,500 km in tail direction. (Courtesy of S. M. Larson, University of Arizona)

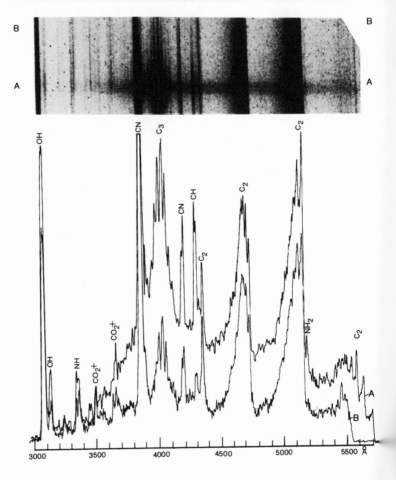

doubt that comets to be discovered in future years will add equally to cometary physics. In addition, new observational techniques applied to both known and newly discovered comets hold great promise for the future.

The material in this chapter is based on very recent research. The ideas are not yet proven, but they are worthy of discussion. The reader should keep abreast of up-to-date journals for discussion, criticism, and evolution of the ideas presented.

8   *Missions to comets*

In our discussion to this point, we have talked about much that has been learned about comets from ground-based and earth-orbiting observatories. And yet, there remain many perplexing mysteries that seem to be beyond our grasp. Solving these problems will probably require us to take a close-up look at a number of comets. The 1980s may well see the first space mission for the in situ exploration of a comet. Many vitally important discoveries will undoubtedly result from such a mission. For example, reconnaissance of a comet may result in (1) determination of the nature, and indeed the existence, of the nucleus by landing on it or by producing a photograph; (2) determination of the physics of the cometary atmosphere, including its composition, density, and motions; (3) determination of the nature of the solar wind–cometary interaction; (4) determination of the mechanism responsible for the molecular ions that dominate comet tails; (5) measurement of the magnetic fields and possible currents in the ion tails; and (6) determination of the properties of the cometary dust by measurements in the coma or dust tail.

The scientific reasons for a comet mission are obvious and should make the whole idea easily salable. Recall the summary of fascinating topics relating to the origin of comets and the solar system given in Chapter 5, and the variety of chemical and physical processes discussed throughout this text. The prime question is: When should the mission fly? The expected apparition of Halley's comet in 1985–6 provides an answer to the question. Halley's comet is an obvious prime target. Halley is a relatively young, large comet that exhibits considerable activity near the sun. Thus, it is scientifically interesting. And, as a bright, famous comet, it will generate considerable public interest in the mission and its outcome.

There are a number of scientific and mission-related guidelines that must be considered in final target selection. These have been suggested by Farquhar in his paper entitled "Mission Strategy for Cometary Exploration in the 1980s," and we quote them here. Bear in mind that Farquhar's interest at the time was in flyby missions, not rendezvous.

The most important guidelines from the scientific viewpoint are:

1. The mission set should be made up of different types of comets. For example, both gaseous and dusty comets should be represented. A comet that has displayed physical characteristics associated with long-period comets should also be included (Halley is the logical choice).

196

2. Comets with a long history of prior observations are preferred. Spectroscopic measurements are particularly useful. [Farquhar, 1974:3]

The most important guidelines from the mission-related viewpoint are:

1. *Relative velocity at encounter.* A small flyby speed will maximize the time available for in situ measurements of the cometary atmosphere, reduce smear in imaging experiments, and minimize the probability of neutral-molecule impact fragmentation.
2. *Targeting errors at encounter.* A sufficiently small miss distance is essential for adequate science return from the imaging and mass spectrometer experiments.
3. *Launch energy requirements* (C_3). Total mission cost is directly related to the launch energy requirement. Small values of C_3 permit the use of smaller and less expensive launch vehicles.
4. *Heliocentric distance at encounter.* Comets are generally more active at smaller heliocentric distances.
5. *Geocentric distance at encounter.* Data rates are higher for smaller earth distances.
6. *Encounter geometry.* Cross-sectional mapping of the cometary atmosphere is preferred.
7. *Earth-based sighting conditions before and during encounter.* Adequate dark time is required to ensure effective ground-based observational support. Recovery should occur at least three months before encounter. [Farquhar, 1974:4]

Most of the mission-related points made by Farquhar are self-explanatory. In item 3 the quantity C_3 (km^2 sec^{-2}) is the launch energy per unit mass, which is a function of the orbit required for the mission and the time of launch. In item 5 the reference to data rates reflects the fact that the scientific data from the experiments are telemetered to earth. For a first direct mission, return of neither a cometary sample nor of the spacecraft itself is being contemplated.

In addition to the scientific and mission-related guidelines as noted above, there are serious programmatic guidelines that influence, and in many cases dictate, how and when direct exploration of comets is carried out. Funding must be forthcoming in the proper amount and in a timely fashion. There are basically three program phases that must be considered: the development of the package of scientific instruments, the conduct of the spaceflight mission itself, and the subsequent detailed analysis of the data. Additional missions should not be carried out until the data from prior missions have been analyzed and understood.

The sum total of these criteria makes it immediately clear that opportunities for comet missions are quite rare. We have already missed an early opportunity to study Encke's comet in 1980. However, that comet is again a possibility in 1990. Other possibilities, in addition to

Comet Halley in 1985–6, are Comet Giacobini–Zinner in 1985, Comet Borrelly in 1987, and Comet Tempel 2 in 1988.

Ballistic flyby missions

The first concept for cometary missions was ballistic intercepts; that is, except for possible mid-course corrections, no propulsion systems were to be used after the vehicle was sent from the vicinity of earth. Thus, these missions were simple flybys. Rendezvous missions in which the spacecraft flies along with the comet for extensive periods require a spacecraft with a propulsion module such as solar–electric propulsion[1] or a suitable chemical stage. Mission concepts have evolved to the use of a propulsion stage. Let us briefly look at ballistic mission possibilities, however.

The orbital elements and orbital geometry for comets Giacobini–Zinner and Borrelly are shown in Figure 8.1; the favorable geometry for both comets near perihelion is clearly shown. The scientific return would be greatly enhanced for each flyby if two probes, carried on the same launch vehicle, are utilized. As shown in Figure 8.2, one probe would pass fairly close to the nucleus on the sunward side, whereas the other probe would pass through the tail region. If only one probe were to be sent, a compromise pass might be through the tail region but rather close to the nucleus on the antisolar side. Targeting of probes is done so that the error ellipse lies one semidiameter outside of the nuclear exclusion zone usually assumed to have a radius of 300 km (to minimize damage to the spacecraft from large dust grains near the nucleus) and has its minor axis along the line of closest approach (to minimize the mass distance). This scheme is shown in Figure 8.3; the impact plane is normal to the relative velocity vector at impact.

The mission to comets Giacobini–Zinner and Borrelly could be accomplished with a single launch and multiple earth swingby maneuvers as shown in Figure 8.4. Note that Figure 8.4 is rendered in a bipolar coordinate system in which the sun–earth line is fixed.

Some sample parameters for the ballistic missions to comets Giacobini–Zinner and Borrelly are shown in Table 8.1. Bear in mind that these are selected typical parameters. The launch energy C_3 for this mission would be 12.3 km^2 sec^{-2}.

Ballistic missions to Halley's comet have also received considerable study. One possible mission is illustrated in Figure 8.5. The major problem with the mission is the high flyby speed of 56.5 km sec^{-1}, which is caused by Halley's retrograde orbit. This high flyby speed is a major problem for neutral mass spectrometer measurements (see below). Even with the fast flyby speed, the time available for measurements is still substantial because of Halley's large size.

Other ballistic missions to Halley's comet have considered a dual launch with two separate intercepts. These missions would provide in situ measurements both before perihelion (say, December 8, 1985) and after perihelion (say, March 20, 1986). Obviously, the comparison of these preperihelion and postperihelion measurements would be of great interest to cometary workers.

Figure 8.1. Orbital geometry and orbital elements for comets Borrelly and Giacobini–Zinner. (Courtesy of R. W. Farquhar, D. P. Muhonen, F. I. Mann, W. H. Wooden, and D. K. Yeomans)

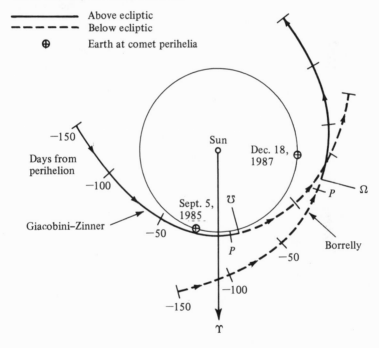

	Giacobini–Zinner		Borrelly
Epoch	1985 Sept. 12.0	Epoch	1987 Dec. 11.0
T	1985 Sept. 5.69583	T	1987 Dec. 18.26830
q	1.0282459 AU	q	1.3567347 AU
e	0.7075535	e	0.6242364
Ω	194.70649°	Ω	74.74641°
ω	172.48534°	ω	353.32292°
i	31.87811°	i	30.32392°

Rendezvous missions

Recently the development of suitable propulsion systems has opened up the possibility of launching a spacecraft to rendezvous with a comet. The spacecraft, as planned, would stationkeep and maneuver in the vicinity of the comet for an extended period of time.

There are basically two rather different propulsion systems capable of carrying out a rendezvous mission. One is the solar sail, which comes in two basic types. The square sail consists of an 800-meter-square sheet of thin plastic film supported by ultralightweight extenda-

Figure 8.2. Possible dual-probe encounter geometry. (Courtesy of R. W. Farquhar, D. P. Muhonen, F. I. Mann, W. H. Wooden, and D. K. Yeomans)

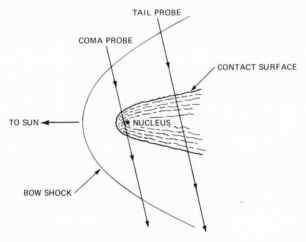

Figure 8.3. Targeting geometry in the impact plane. (Courtesy of R. W. Farquhar, NASA–Goddard Space Flight Center)

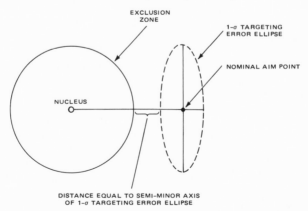

ble booms to serve as the needed masts and spars. An alternate version is the so-called helio gyro (like a pinwheel), some 12,000 meters in diameter, which consists of 8-meter-wide blades. The sail area is essentially the same in both options. The sails are driven by solar radiation pressure. The orbit needed to reach Halley's comet includes an initial approach to within 0.25–0.30 AU of the sun, where many revolutions are used to rotate the orbital plane to coincide with the orbital plane of Halley's comet; this is called a *cranking orbit*. The sail then moves out to rendezvous with the comet some 34 days *after* perihelion at 0.92 AU from the sun. The rendezvous would take place from the antisunward side.

The second basic propulsion system is the ion drive, sometimes called *solar electric propulsion*. Solar energy is utilized via solar cells to generate electric power. The electric power is used to ionize and accelerate massive atoms such as mercury or argon to high exhaust velocities of order 10 km sec^{-1}. The thrust of such an engine is tiny – a few hundredths of a pound. However, the engine can operate continuously for several years on a moderate quantity of propellant. Thus the total

Figure 8.4. Spacecraft trajectory for multiple-encounter mission. (Courtesy of R. W. Farquhar, D. P. Muhonen, F. I. Mann, W. H. Wooden, and D. K. Yeomans)

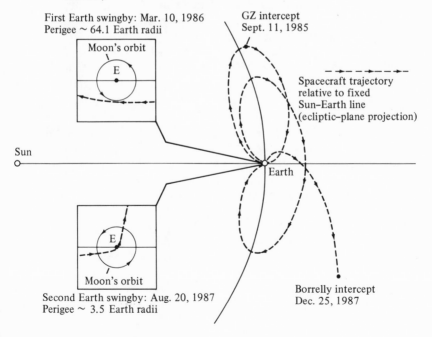

impulse – defined as the integral $\int F\,dt$ extended over the period of engine firing – can be significant for an ion drive engine.

Comet missions now contemplated would require six to eight individual ion engines. The total solar power required for the complement of engines exceeds 25 kw. The solar panels would be a pair of giant wings, each with deployed dimensions of (say) 4 by 35 meters.

The original hope for a rendezvous mission was with Comet Halley. The mission studied included rendezvous with Halley's comet some 50 days before perihelion and continuous monitoring through perihelion and out to a distance of 3 to 4 AU. However, a launch date in May or June 1982 was required, and this in turn required early approval and funding of the mission. Unfortunately, the approval was not forthcoming and an alternate approach was needed.

The celestial mechanicians have discovered that, with the ion drive, it is possible to carry out a dual comet mission. This unique mission would consist of a flyby of Comet Halley and a rendezvous with Comet Tempel 2 (see Figure 8.6). Tempel 2 is a significantly different sort of comet than Halley. It is a member of Jupiter's family of comets, with a period of 5.3 yr. It is smaller than Halley and does not show the high level of activity that characterizes the latter body. The dual comet mission would spend a long period of time in the vicinity of Tempel 2. The Halley flyby would be brief but would permit a probe to be dropped into the vicinity of the comet's nucleus. An independently launched tail probe may also be a possibility. Table 8.2 lists the events that will take place during the mission.

The successful completion of a rendezvous mission with Tempel 2 combined with the Halley flyby and probe could revolutionize the study of comets. The rendezvous will permit detailed studies of the structure of Tempel 2, and the long-term stay in the vicinity of the comet will permit a study of temporal variations in cometary phenomena. In addition, the nucleus could be photographed in some detail, and a landing on the nucleus might be attempted as a terminal maneuver of the mission as the cometary activity subsides with increasing heliocentric distance. Clearly, the possibilities are exciting and the scope of cometary physics could be greatly enlarged.

Table 8.1. *Sample ballistic missions*

Comet	Arrival	Flyby speed (km sec^{-1})
Giacobini–Zinner	September 11, 1985	20.6
Borrelly	December 25, 1987	17.3

At the same time, the rendezvous possibilities greatly enlarge the scientific community that could profit by the mission. If the studies are expanded to include the nucleus and the relatively dense medium of gas and dust close to it in addition to the outer atmosphere and the tail, the rendezvous mission becomes a *total* cometary mission of interest to geologists, geochemists, interstellar astronomers, cosmogonists, and so on, as well as cometary astronomers and solar wind physicists. Expen-

Figure 8.5. Possible Halley ballistic intercept in 1985. The orbit returns the spacecraft to the earth's vicinity 1 year after launch. The advantage of this "boomerang" trajectory concept is retargeting the spacecraft to another comet after the Halley flyby; the idea is illustrated in Figure 8.4. (Courtesy of R. W. Farquhar and W. H. Wooden, NASA–Goddard Space Flight Center)

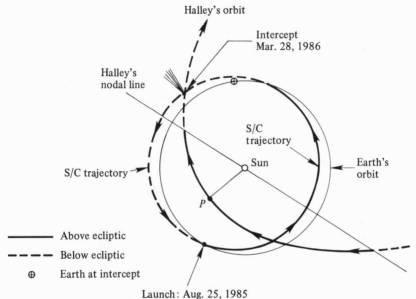

Launch: Aug. 25, 1985
Earth return: Aug. 25, 1986

Encounter parameters at Halley

Intercept date	Mar.28, 1986 ($P + 47$ days)
Sun distance	1.13 AU
Earth distance	0.59 AU
Phase angle	113.8°
Flyby speed	58.4 km sec^{-1}

Launch parameters		Spacecraft trajectory	
Launch energy - C_3	36.3 km^2 sec^{-2}	Perihelion	0.82 AU
Declination of		Aphelion	1.18 AU
launch asymptote	38.9°	Inclination	6.7°
		Period	1.00 year

sive missions such as the dual comet mission require widespread support in the scientific community for approval and funding. Public support for such a mission can also be important. This mission clearly ranks high in both areas.

The excitement of the possible rendezvous mission has been tempered by a major problem: The ion drive propulsion system will not be available in time. Thus, at the time of this writing in late 1980, it is clear that NASA will not fly the Halley flyby–Tempel 2 rendezvous. There still is a possibility for a ballistic intercept mission to Halley, a mission called the *Halley Intercept*. Several other nations have picked up the ball on Halley missions. The European Space Agency is planning a Halley mission – aptly named *Giotto* (see Chapter 10). Both Japan and the Soviet Union also have plans for a mission.

Experiment package

The detailed experiment package to be carried on a comet mission is not known at this writing. NASA's normal procedure is to issue an announcement of opportunity (AO) listing the mission con-

Figure 8.6. The Halley–Tempel 2 mission. (NASA–Jet Propulsion Laboratory)

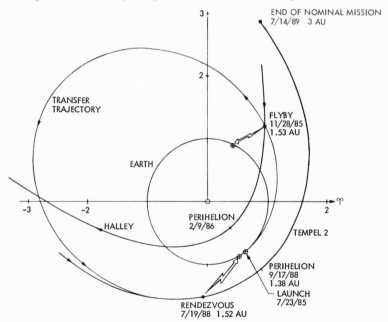

Heliocentric Trajectory (Ecliptic Plane Projection)

straints (mass, power, volume, noninterference, etc.) and to request proposals from the scientific community. These proposals are evaluated (usually by a committee of peer scientists), and the final selection is made by the appropriate officials at NASA headquarters.

Nevertheless, the broad outlines of the experiment package seem clear given the scientific objectives.

1. Imaging system. The imaging of the cometary nucleus could be carried out from a stable platform or through the use of a "spin-scan camera" if the spacecraft is spinning. If the latter is not feasible and a spinning spacecraft is indicated, provision must be made for a despun platform. Several of the instruments described below would also be mounted on such a platform. Studies have shown that an imaging device would see the nucleus and that the so-called false nucleus would not be a problem. We do not see the typical cometary nucleus from earth solely because of insufficient resolution.

2. Ion mass spectrometer. The large range of energies (~ 0.1 eV to ~ 1 keV) and the large range of possible masses (1 to 60 atomic mass units or more) are difficulties for the usual ion mass spectrometers. Possible solutions are to fly several mass spectrometers with different ranges or to use types with a wider range. Additional development is needed here.

Table 8.2. *Halley–Tempel 2 mission events (nominal)*

Launch	
Date	August 1, 1985
C_3	35 km^2 sec^{-2} (max)
Halley flyby	
Date	November 28, 1985
Mission time	120 days
Time before perihelion	73 days
Solar distance	1.5 AU
Tempel 2 rendezvous	
Date	July 18, 1988
Mission time	1080 days
Time before perihelion	60 days
Solar distance	1.5 AU
End of mission	
Date	July 14, 1989
Mission time	1440 days
Time after perihelion	300 days
Solar distance	3.0 AU

3. Neutral mass spectrometer. The neutral species must be ionized
    before striking any surface, and the usual technique is to use a
    crossed electron beam. The principal problem is the very low
    efficiency of ionization (~1%). Additional development is
    needed. Devices using so-called field ionization techniques
    may be the answer.
4. Dust composition device. This measurement is very difficult
    and very important. The dust particles need to be vaporized
    and then analyzed by means of a mass spectrometer. A
    major development program is needed here.
5. Nuclear radiometer. This item is intended to measure the
    temperature of the surface layers. Traditionally, an infrared
    device would be used, but a millimeter-wave radiometer may
    have advantages for the cometary environment, as well as
    providing measurements at different depths.
6. Ultraviolet spectrometer
7. Lyman α photometer
8. Magnetometer
9. Electron analyzer

Figure 8.7. Artist's schematic of the Halley–Tempel 2 mission. (NASA–Jet
Propulsion Laboratory)

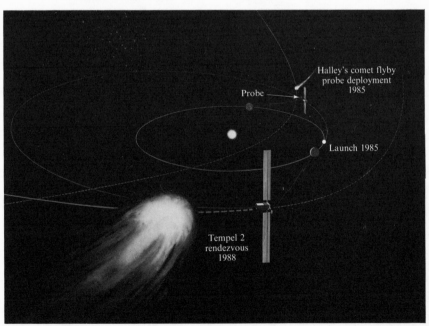

10. Plasma analyzer
11. Plasma wave detector
12. Dust detector

Items 6 through 12 can be regarded as basically developed instrumentation available for flight opportunities with a relatively minor development effort to modify or adapt them for specific cometary missions.

A necessary facet of any mission to a comet should be a comprehensive and vigorous program of supporting ground-based observations. In addition, theoretical comet scientists must also be involved.

A dual comet mission such as the Halley flyby–Tempel 2 rendezvous (Figure 8.7) with an instrument package such as we have described above would be truly revolutionary. However, if political realities prevent that particular mission, the scientific community must persist, even though it would be the *third* major lost opportunity. It is tremendously exciting to contemplate the fascinating discoveries awaiting completion of a cometary mission of almost any description. Unfortunately, new bright comets are rarely discovered in time to mount a full-scale mission. So, we will have to look to periodic comets that will return in later years. The prospective payoff is too great to give up in frustration.

Part IV

The lay perspective

Bright comets tend to catch the public fancy. Unfortunately, a great deal of nonsense has been written to pander to this interest. We should try to counter such exploitative literature by presenting the modern scientific picture of comets to the public.

In Part IV we will take a somewhat lighter look at comets. We will first address the question: Are comets hazardous to your health? Finally, we will take a very brief look at comet lore.

Facsimile headlines showing *New York Times* coverage of Halley's comet on May 10, 16, and 18, 1910.

BARNARD PICTURES OF HALLEY'S COMET

Taken at Yerkes Observatory May 4, They Tally with Observation from Times Tower May 5.

VIEWED BY MISS PROCTOR

Negatives Show the Tail Extending 20 Degrees, Equivalent to 24,000,000 Miles in Length.

IN COMET'S TAIL ON WEDNESDAY

European and American Astronomers Agree the Earth Will Not Suffer in the Passage.

TELL THE TIMES ABOUT IT

And of Proposed Observations— Yerkes Observatory to Use Balloons if the Weather's Cloudy.

TAIL 46,000,000 MILES LONG?

Scarfed in a Filmy Bit of It, We'll Whirl On In Our Dance Through Space, Unharmed, and, Most of Us, Unheeding.

SIX HOURS TO-NIGHT IN THE COMET'S TAIL

Few New Yorkers Likely to Know It by Ocular Demonstration, for It May Be Cloudy.

OUR MILLION-MILE JOURNEY

Takes Us Through 48 Trillion Cubic Miles of the Tail, Weighing All Told Half an Ounce!

BALLOON TRIP TO VIEW COMET.

Aeronaut Harmon Invites College Deans to Join Him in Ascension.

MAY SEE COMET TO-DAY.

Harvard Observers Think It May Be Visible in Afternoon.

MAY BE METEORIC SHOWERS.

Prof. Hall Doubts This, Though, but There's No Danger, Anyway.

YERKES OBSERVATORY READY.

Experts and a Battery of Cameras and Telescopes Already Prepared.

CHICAGO IS TERRIFIED.

Women Are Stopping Up Doors and Windows to Keep Out Cyanogen.

9 *Are comets dangerous?*

There are two areas in which comets, at least in principle, pose a real danger. One need only look at the list of substances found in the comas and tails of comets to see that they are highly toxic. The most dangerous constituent is cyanogen (CN), a poisonous, flammable, and colorless gas. Because comets typically move in highly eccentric orbits and many have perihelia inside 1 AU, there exists a finite but small probability of a near miss or collision with the earth. If a near miss occurs, the atmosphere could be polluted with toxic gas. If a collision occurs, the destruction could be considerable. The kinetic energy of an object striking the earth at parabolic velocity exceeds by over two orders of magnitude the chemical energy in a comparable mass of TNT. How real are these dangers? To find out, we must look at evidence.

Poisonous gases

The earth passed through the tail of Halley's comet in May 1910. Needless to say, nothing serious happened (see the figure on the Part IV opening page). The amount of cyanogen was, in fact, completely inconsequential. Nevertheless, a great deal of fear was aroused in the general public, and the doomsayers were predicting the end of the world. Charlatans appeared on the scene selling pills and mouth inhalers that allegedly would provide protection against the lethal gases. The danger from poisonous gases is not, in fact, real. A relatively simple calculation shows that the amount of poisonous gas we might receive from a close encounter with a comet would be very small. Let's assume that all the gas in a comet is noxious, although this is certainly wrong, as our earlier discussions have shown. Most of the gas in a comet is within one scale height from the nucleus. We will be very generous and assume the scale height to be 10 times the nuclear radius, r. That estimate is clearly excessive. Then the mass of the cometary gas is very roughly

$$M_{at,c} \sim 4\pi r^2 \cdot (10r) \cdot N \cdot 30 \times m_H \qquad (9.1)$$

where N is the number surface density of gas in the nucleus, taken to be 10^{14} cm^{-3}, m_H is the mass of a hydrogen atom, and 30 is the (overestimated) mean molecular weight. With $r = 10$ km or 10^6 cm, we find the upper limit of the mass of the cometary atmosphere to be $M_{at,c} \sim 10^{12}$ g. This number should be compared with the estimated mass of the earth's atmosphere, $M_{at,E} = 5 \times 10^{21}$ g. Thus, noxious gas from a comet could add at most 1 atom in 10^{10} to the atmosphere by a close encounter. Incidentally, the density $N = 10^{14}$ cm^{-3} at the cometary nucleus corresponds to the atmospheric density at 85 km above the earth's surface. Thus, the cometary gas could not penetrate far into the earth's atmosphere even if an unlikely encounter should occur. The danger from poisonous cometary gases is truly nil.

Cometary germs

Hoyle and Wickramasinghe (1978a) hypothesize that the building blocks of life in the universe have developed from the organic molecules found in interstellar space. They argue that, as clumps of interstellar grains grow, complex prebiotic molecules form within the grains. Some mechanism, such as the formation of cell walls, protects the molecules from disruption by ultraviolet light in interstellar space. The authors point to the discovery of biologically important amino acids in carbonaceous chondrites as evidence for their point of view.

The authors then carry their argument one step further. Comets formed within these interstellar clouds contain the grain clumps and their prebiotic material. As time goes on, primitive life forms could evolve within the comet. As comets are captured by planetary systems, they spread life throughout the universe.

Hoyle and Wickramasinghe interpret their theory to mean that comets are spreading germs throughout the universe. In an article in *New Scientist* (1978b), they analyze the attacks of influenza in some schools in Britain. The attacks are characterized by a very rapid spread over a number of widespread areas, despite evidence that the virus takes a long time to travel a short distance overland. The authors conclude that the influenza virus descends through the atmosphere and settles at ground level, where it exhibits a fine-scale patchiness. Those in contact with the virus would be exposed at more or less the same time. The authors are continuing their research in this area, even looking at a possible extraterrestrial origin for the plagues of the Middle Ages.

These ideas are very controversial. Nevertheless, the authors are distinguished scientists who have put forward their ideas in a clear, coherent form. The whole problem of the origin of life in the universe is one that has received renewed attention recently, and the premises of the Hoyle–Wickramasinghe theory will be carefully scrutinized in the future.

Sherwood Chang (1979) of the NASA–Ames Research Center, Extraterrestrial Research Division, views the study of comets as an important link in the study of the origin of the earth, its atmosphere, and life itself. An important step in the study is to ascertain whether comets condensed in interstellar clouds, as Hoyle and Wickramasinghe (1978a) suggest, or were formed in the solar nebula as a by-product of the formation of the solar system. In either case, comets do carry within themselves organic molecules that are important for the creation of life. Have comets played a major role in the origin of life? Just the fact that we can seriously ask this question points to the extreme urgency of a vigorous program of cometary research. We do not yet have the information we need to understand the role of comets in the history of the early solar system. We must make every attempt possible to obtain that information.

Cometary impact

The second problem is real enough, although rather improbable: The earth could be struck by a comet. This possibility has caused a great deal of alarm, particularly when sparked by an apparently respectable calculation. For example, the February 15, 1872, issue of *Nature* notes:

We have reason to know that many weak people have been alarmed, and many still weaker people made positively ill, by an announcement which has appeared in almost all the newspapers, to the effect that Prof. Plantamour, of Geneva, has discovered a comet of immense size, which is to "collide" as our American friends would say, with our planet on the 12th of August next. [*Nature* 5:310]

The writer goes on to point out that the masses of comets are very small – essentially they are "wind-bags" – and, hence, little damage would be expected. The item concludes with some advice to the reader that is appropriate for most fears associated with comets, namely that he make financial "offerings" to the Royal Astronomical Society. In this way "Science would be benefited, and, doubtless, the omen would be averted – at all events they always have been."

There is one location on earth that may have been the scene of a cometary collision. A little after 7 A.M. local time on June 30, 1908, in central Siberia (about 1000 km north of Irkutsk) a gigantic explosion occurred and wrought havoc in the countryside. Searing was traceable for about 18 km, and the general devastation extended for a radius of 30 to 40 km. Perhaps the most graphic evidence was the large number of trees blown down with their trunks pointing radially away from the center of the explosion (Figure 9.1). The sound was heard at distances over 1000 km. The event produced an earthquake, and a seismic wave was recorded. Magnetic disturbances that resemble the events known to occur after nuclear explosions were also recorded.

The explosion was preceded by a very bright fireball that crossed from the southeast to the northwest and left a thick dust train. Flames and a cloud of smoke were seen over the site, and incandescent matter was thrown to heights of 20 km. The nights after the explosion were exceptionally bright in western Asia and Europe. The amount of dust in the atmosphere was sufficient to cause a reduction in the transparency of the atmosphere as measured 2 weeks later in California.

Expeditions to the sparsely populated region did not begin until 19 years later in 1927. Then an expedition headed by L. A. Kulik started the scientific study of the Tunguska area. Besides establishing the general description of the area and the event noted above, the expedition found an abundance of magnetite and silicate globules (diameters 5–450 μm) in the soil. An aerial photographic survey of the region in 1938–9 was important in establishing some of the results cited.

Figure 9.1. The forest felled by the Tunguska event some 8 km from the center of the explosion. (Sovfoto)

A fascinating and relatively poorly observed event such as this one has inspired many proposed explanations. Among these are alien spacecraft, a mass of antimatter, a mass of near-critical fissionable material, and a black hole. These exotic explanations were largely inspired by some apparent deficiencies in the most likely theory, namely, that the earth was struck by a small comet; the basic ideas were proposed by Whipple in 1930 and Astapovich in 1933. There has been considerable recent work on the Tunguska event.

Comparisons of the old seismic and acoustic records with similar records of contemporary nuclear air explosions lead to the conclusion that the Tunguska explosion had an energy of $5 \pm 1 \times 10^{23}$ ergs and occurred at an altitude of 8.5 km. If the energy of the explosion was derived entirely from the kinetic energy of the incident mass, that is, if the energy from chemical reactions can be neglected, then the size of the body can be deduced. The geocentric speed of the body has been estimated at 47 km sec^{-1}. Thus, the comet's mass was approximately 5×10^{10} g. For an assumed density of 1.14 g cm^{-3}, the cometary nucleus would be about 40 meters in diameter.

Standard expressions for the brightness of a comet can be used to estimate apparent brightness of the comet at various times before impact. Because of the small diameter (which is an order of magnitude or more smaller than most values in the literature), one would expect that the comet would be quite close to earth before becoming visible. Brown and Hughes (1977) have calculated that it would have been brighter than +11th magnitude, the value at which comets were discovered around the turn of the century, for only about 1 hr. It would have been visible to the naked eye (brighter than +6th magnitude) for only about 3 min. Calculations by Fesenkov (1966) show that all possible orbits calculated for the fireball have a small angular separation from the sun. This fact and the fact that it came from a point in the dawn sky in the Tunguska region would have made this body very difficult to detect. Hence, the nonobservance of the comet is no mystery at all; in retrospect, it should have been expected.

There was also an enhancement in the ^{14}C production as measured in 1909, and this too has been attributed to the Tunguska event. Apparent lack of a mechanism to produce substantial amounts of ^{14}C in a cometary impact has been regarded as a deficiency, but this may not be so. Brown and Hughes (1977) have shown that the number of neutrons ($\sim 10^{27}$) required to produce the enhanced ^{14}C abundance is entirely compatible with a rough scaling comparison with results from solar flares. Obviously, the comparison is rough, but clearly the ^{14}C anomaly cannot be considered a problem of the cometary impact hypothesis. As we described above, neither can the lack of discovery of the cometary body.

Thus, the evidence available provides support for the hypothesis that the Tunguska explosion and associated events were produced by the impact of a small comet. Brown and Hughes (1977) note that "nothing more exotic need be involved." Most of the phenomena described above follow naturally from the impact hypothesis. The small globules found in the soil were molten "drops" of cometary dust that solidified as they fell to earth. Finally, the enhanced night sky luminescence seems simply explicable only on the cometary impact hypothesis. If the comet had a gas or dust tail, it should have pointed directly away from the sun and would have encountered the atmosphere in the regions of high sky brightness. The night sky brightness, which was 50–100 times the normal value, could be produced by dissipation of the tail in the atmosphere.

The Czech astronomer Lubor Kresak has studied the available evidence concerning the Tunguska event and concluded that it was a β Taurid meteor. This meteor shower has been associated with Encke's comet. Kresak asserts that the meteor was a particularly large, inactive chunk of that comet.

Now, is this magnificent event sufficient reason to fear comets? The original event occurred in an isolated area sparsely populated by scattered bands of nomadic reindeer herders, and there was no large loss of life. Clearly, if such an event were to occur over a major city, we would have a disaster. Assessment of the situation requires discussion of the odds. Estimates of the flux of cometary bodies of various sizes [based on Everhart's (1972) calculations] indicate that the chances of a similar body striking the earth are on the average once every 2000 years. In addition, the devastated area is fairly local. For a radius of 40 km, the 5000 km^2 area is about 0.001% of the earth's total surface area. Thus, the event is rare and only 10^{-5} of the earth's surface is affected. The danger is real enough, but the probability of being directly affected by the impact of a comet is extremely small. No one should lose much sleep over this situation even if agitated by uninformed doomsayers.

10   *Comet lore*

A very bright comet is a beautiful sight to behold. Because a comet is brightest when it is simultaneously near perihelion and near the earth, a naked-eye comet is most often seen in the western sky just after sunset or in the eastern sky just before sunrise. Then, the comet can be the most obvious object in the sky. Naturally enough, an impressive comet will excite considerable public interest. Both of the authors have spent considerable time speaking to general audiences and answering questions from them, as well as from the press. People are genuinely interested in the nature of comets – probably because those bright enough to catch the public interest are infrequent.

Recent comets have brought home an unfortunate truth: Most people will not be able to see even bright comets from their home sites in the future. Comet Kohoutek (1973f) was discovered many months before it passed perihelion. Preliminary calculations indicated that it would be very bright during the Christmas season of 1973–4, and considerable excitement was built up beforehand. After the fact, the press dubbed Comet Kohoutek a dud; most people never saw it. Yet to astronomers on their mountaintop observatories, it was a conspicuous naked-eye object for almost a month. Part of the difference was due to bad weather conditions in the populous northeastern United States during the time Kohoutek was visible. But a big problem was (and is) pollution. Filthy air and bright city lights conspire to overwhelm the subtle light of a comet. Let us hope that this situation will change in the future.

Although much interest in comets comes from a public seeking knowledge about a fascinating phenomenon, not all interest is so nobly motivated. Superstition, fear, and charlatanism rear their ugly heads, too. Unhappily, the idea that comets are omens is widespread. As we saw in Chapter 9, a real danger from comets does exist, though it is very small. We have already discussed some of the early conceptions of comets in Chapter 1. Here we will look more closely at some of the superstitions.

Comets and superstition

Will Durant (1954:80) points out that "astrology antedated – and perhaps will survive – astronomy; simple souls are more interested in telling futures than in telling time." The belief that comets are bad omens is an integral part of Aristotle's philosophy. However, it is not difficult to believe that the concept is far more ancient. Memorable events in human existence occur constantly. A bright comet might have appeared to Stone Age people at the time of a great drought when food was scarce, or at the time of a particularly disastrous hunt when tribesmen were injured or killed. The next bright comet would strike fear in their hearts. And the fear would be borne out by another setback in the people's hard existence. By extension, it is not difficult to imagine why similar superstitions continue in regions where civilization has barely intruded.

That superstitions regarding comets are still touted in the modern civilized world is more difficult to understand. It is clear that some writers are exploiting the belief for their own purposes. It is difficult to imagine a physical mechanism whereby a passing comet could cause specific events such as the death of a king, and scientists do not take the idea seriously.

The Roman Catholic Church was one of the bastions of cometary superstition during the Middle Ages, and the idea persisted in teachings through the Reformation period. A scriptural basis for the belief was the book of Joel, Chapter 2, verses 30 and 31: "And I will shew wonders in the heavens and in the earth, blood, and fire, and pillars of smoke. The sun shall be turned into darkness, and the moon into blood, before the great and terrible day of the Lord come." Bright comets were warnings to keep us on our toes. St. Thomas Aquinas' great *Summa Theologica,* written in the thirteenth century, is full of references to the influences of heavenly bodies on human events. With scripture and the church's greatest thinkers behind the concept, it is no surprise that the idea has persisted.

Modern superstition is not a high point of humanity's intellectual development, and it should not be dignified by extensive discussion. The spirit of the historical attitude is captured in Figures 10.1, 10.2, and 10.3, and by the following quotation from Sir Stanislaus Lubeinietski in his *Theatrum Cometrium,* published in 1668: "Known to God and man . . . never had there been a disaster without a comet or a comet without a disaster . . . had there not been a comet overhead which had been responsible for the epidemic of sneezing sickness among the cats of the Rhenish areas of Westphalia?" (Haney 1965:177). If human misfortune truly were connected with the appearance of comets, the sky should be bright with them!

It is relatively easy to forgive the colorful folklore of previous centuries, and we can be amused with Mark Twain's statement that he came (born 1835) and would go (died 1910) with Halley's comet. However, it is difficult to consider seriously the following statement pub-

Figure 10.1. Title page of John Hill's *Alarm to Europe*. (By permission of the Houghton Library, Harvard University)

A N
A L L A R M
T O
E U R O P E :
By a Late Prodigious
C O M E T
feen November and December, 1680.

With a Predictive Difcourfe. Together with fome preceding and fome fucceeding Caufes of its fad Effects to the *Eaff* and *North Eaftern* parts of the World.

Namely, *ENGLAND*, *SCOTLAND*, *IRELAND*, *FRANCE*, *SPAIN*, *HOLLAND*, *GERMANY*, *ITALY*, and many other places.

By *John Hill* Phyfitian and Aftrologer.

The Form of the *COMET* with its Blaze or Stream as it was feen *December* the 24th. Anno 1680. In the Evening.

London Printed by *H. Brugis* for *William Thackery* at the Angel in Duck-Lane.

8 Jan. 168½

lished in this decade: "Every comet in history has been associated with war, catastrophe, and extraordinary events . . . Like it or not, Kohoutek will be, too." (Goodavage 1973:1). To be fair, the author of the same book warns the reader not to "be fooled by weighty scientific pronouncements." The cover of a Children of God pamphlet in 1973 entitled *What Will the "Christmas Monster" Bring?* captures the spirit of the situation quite well.

Clearly, the only effective counter to this kind of argument is education, and public education is an obligation of the cometary scientist.

Comets and literature

Most references to comets in literature are based on the folkloric view of them as omens or producers of disasters. Examples are William Shakespeare in *Julius Caesar:*

Figure 10.2. French cartoon reflecting fears that a comet in 1857 could collide with the earth.

Figure 10.3. Title page of Increase Mather's *Discourse Concerning Comets.*

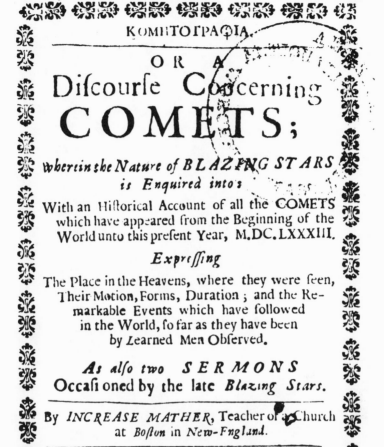

ΚΟΜΗΤΟΓΡΑΦΙΑ.

OR A

Difcourfe Concerning

COMETS;

Wherein the Nature of *BLAZING STARS*
is Enquired into:

With an Hiftorical Account of all the COMETS
which have appeared from the Beginning of the
World unto this prefent Year, M.DC.LXXXIII.

Expreffing

The Place in the Heavens, where they were feen,
Their Motion, Forms, Duration; and the Remarkable Events which have followed
in the World, fo far as they have been
by Learned Men Obferved.

As alfo two *SERMONS*
Occafioned by the late *Blazing Stars.*

By *INCREASE MATHER*, Teacher of a Church
at *Bofton* in *New-Fngland*.

Pfal. 111. 2. *The works of the Lord are great, fought
out of all them that have pleafure therein.*
Amos 9. 6. *He buildeth his ftories in the Heaven.*

BOSTON IN NEW-FNGLAND.
Printed by S. G. for S. S. And fold by *J. Browning*
At the corner of the Prifon Lane next the Town-
Houfe 1683.

Figure 10.4. Comet-related cartoons from issues of *Punch;* Top, June 1, 1910; bottom, December 5, 1906.

Burglar (with sudden enthusiasm for astronomy). "'SCUSE ME, GUV'NER, CAN YOU TELL ME WHERE I CAN GET A VIEW OF THIS 'ERE COMET?"

OUR UNTRUSTWORTHY ARTIST IN LONDON.
DISCOVERY OF A COMET AT GREENWICH OBSERVATORY.

When beggars die, there are no comets seen,
The Heavens themselves blaze for the death of Princes.

John Milton in *Paradise Lost:*

Satan stood
Unterrified, and like a Comet burn'd
That fires the length of Ophiuchus huge
In th' Artick sky, and from its horried hair
Shakes pestilence and war.

Shakespeare in *Henry IV* wrote these words, which sound much better to our ears:

By being, seldom seen, I could not stir,
But, like a comet, I was wondered at.

Some departures from the folklore mold were made in the prose of two famous science fiction writers. *Off on a Comet* was published by Jules Verne in the 1860s. It is a good adventure story (given a few liberties), and it has an interesting scientific curiosity. The comet of the story has a satellite, and this fact allows the astronomers to calculate the mass of the comet. Recall that no real comet has had its mass directly determined as of this writing.

H. G. Wells has an interesting twist on the folkloric approach. In *In the Days of the Comet* he describes a comet with a sunward tail and an unprecedented band in the green. The fictional comet approaches the earth, but the influences on earth and its inhabitants are all favorable.

A fairly recent entry in the literature of comets is the book *Lucifer's Hammer* by Larry Niven and Jerry Pournelle (1977). It is a fictional account of a collision between the earth and a very large comet. It is particularly noteworthy here because the authors have taken care to be as scientifically accurate as they can, consistent with their tale.

Comets have often been a favorite topic for humorous cartoons. Two of our favorites appeared in *Punch.* One was associated with the 1910 apparition of Halley's comet (see Figure 10.4, top). A masterpiece of the cartoonist's art is shown in Figure 10.4 (bottom): The Greenwich Observatory buildings are faithfully represented, but almost all else is delightful fantasy.

Comets in art

The Bayeux Tapestry (see Figure 1.2), which was completed in the eleventh century reputedly by Queen Matilda, wife of William the Conqueror, is certainly one of the most famous works depicting Halley's comet. The comet is shown as an omen for the outcome of the Battle of Hastings.

Halley's comet reached perihelion in October 1301. A fresco

Figure 10.5. Giotto's *Adoration of the Magi*. Halley's comet at its appearance in 1301 probably served as the model for the comet shown at the top of the fresco. (Alinari/Editorial Photocolor Archives)

Figure 10.6. Halley's comet in its 1145 apparition was probably the model for the comet shown on the *Eadwine Psalter*. (Courtesy of the Master and Fellows of Trinity College, Cambridge University)

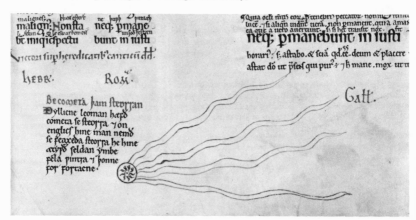

painted by Giotto di Bondone (1267–1337) around 1303 shows a typical nativity scene, but the star of Bethlehem looks remarkably cometlike (Figure 10.5). Was Giotto so impressed with Halley's comet that he used it to represent the Christmas star? Some art historians think so. Certainly the rendition of the comet agrees well with several written descriptions. It does look surprisingly realistic.

Art historian Roberta J. M. Olson (1979) has studied this and other portraits (Figure 10.6) of Halley's comet produced over the years. The

Figure 10.7. Albrecht Dürer's *Melencolia I* of 1514. (Courtesy of the Metropolitan Museum of Art, Dick Fund, 1943)

earliest that she has found represents the A.D. 684 apparition, and it appears in the *Nuremberg Chronicles* published only in 1493 (Figure 1.1). The charming woodcut is supported by a text that recounts all the disasters brought on by the comet. Albrecht Dürer, in a beautiful engraving entitled *Melencolia I,* dated 1514, presented (Figure 10.7) a "luminous and dynamic comet" that was "much less naturalistic than Giotto's." Once again, the picture seems to express the unhappy side of cometary lore.

Many representations of comets dwell on the disasters they are thought to foretell. We should remember Giotto's fresco. There Halley's comet is associated with an event that many people see as one of great joy and promise.

In the same spirit of optimism we would like to conclude this discussion with the hope that each step forward in unraveling the mystery of comets (or any other natural phenomenon) brings great pleasure to all who look to the sky as a source of beauty and intellectual challenge.

Notes

Chapter 1. Comets in history

1 Toscanelli was also a confidant of Christopher Columbus.
2 Although Tycho's reputation for accuracy was accepted in his own time, it can be independently verified. His observations were used in the nineteenth century to compute the orbit for the comet of 1577. In addition, the observations in 1577 were relatively crude by Tycho's standards. With better measurements, his positional measurements of the comet of 1585 had a mean error of only 1'.
3 He is generally well known for his charting of the lunar surface.
4 Clairaut understood that unknown planets beyond Saturn could influence his calculations. His methods were correct, and his prediction would have been better still if he had been aware of the existence of Uranus and Neptune and had, as well, better values for the masses of Jupiter and Saturn.
5 Later studies would confirm that Halley's comet had been sighted on many earlier occasions, dating back to perhaps 466 B.C. (Chinese records).
6 Additional unequivocal proof was added by the discovery of Neptune.
7 Olbers is famous for the cosmological paradox (that bears his name) concerning the brightness of the sky at night.
8 A close approach to Mercury in 1835 allowed an accurate determination of the planet's mass.
9 Suggested in the twentieth century.
10 Daniel Kirkwood (1814–95) wrote in 1861: "May not our periodic meteors be the debris of ancient but now disintegrated comets, whose matter has become distributed round their orbits?"
11 In modern terminology, these would be called members of Jupiter's comet family.

Chapter 2. Development of modern ideas on the physics of comets

1 The only qualification must be that the velocities derived for the knots must represent material motion rather than a wave of some sort.
2 These bands were often called *hydrocarbon bands* in the early literature; they have nothing to do with hydrocarbon because they arise from the carbon molecule C_2. The bands have been named in honor of Swan.
3 Henry Draper (1837–82) obtained spectra of the same comet a few days later.
4 Schwarzschild's and Kron's plates have for some time provided the most reliable determination of CO^+ densities in comet tails. The photographs were, in essence, taken by chance, but they were standardized.
5 Öpik has pointed out that the Δ^{-2} law is valid only in the ideal case in which the total brightness of a comet is measured. If brightness below a fixed level is lost in the sky background, the exponent of Δ will be larger, i.e., between -2 and -1.
6 Temperatures of meteoritic blocks 1 cm in diameter have been calculated by M. G. J. Minnaert.

227

7   One can easily show this relationship beginning with the series expansion

$$\ln x = (x - 1) - \tfrac{1}{2}(x - 1)^2 + \cdots \qquad 0 < x \le 2$$

and converting to common logarithms.

8   Lagrange also noted the possibility that comets originated in the explosion of a planet. This idea has been revived recently (Chapter 5).

9   Currently a prize is offered for an amateur discovery of a comet (see *Mercury* **8,** 60).

10   A specific exception, for example, was Comet Humason (1961e), which showed activity and a CO^+ spectrum out to 5 AU.

Chapter 3. Dynamics of comets

1   Renamed by approval of the International Astronomical Union in 1948 from Comet Pons–Coggia–Winnecke–Forbes. Crommelin found that the various discoveries referred to one comet with a period of 28 years. The comet returned to perihelion in 1956 as he had predicted.

Chapter 4. Structure of comets

1   One will often find the exponential terms written, correctly, as $e^{-(r-R)/R_0}$. However, because the equations are usually applied at values of r, large compared to the nuclear radius R, our forms are sufficiently accurate.

2   Note that the *shape* of the curve given by equation 4.32 is symmetrical with respect to the two scale lengths involved.

Chapter 5. Model and origins of comets

1   This does not deny the importance of accretion at some time in the evolution of the nebula. Indeed, the chemistry of some chondrites demands it.

Chapter 8. Missions to comets

1   This is an ion-rocket engine powered by solar energy.

Suggested readings

General works

Alfvén, H., and Arrhenius, G. 1976. *Evolution of the Solar System*. NASA SP-345. Washington, D.C.: GPO.

Arrhenius, G., and Alfvén, H. 1973. *Asteroid and Comet Exploration*. NASA CR-2991. Washington, D.C.: GPO.

Chebotarev, G. A., Kazimirchaka-Polonskaya, E. I., and Marsden, B. G. (eds.). 1972. *The Motion, Evolution of Orbits and Origins of Comets*. IAU Symposium No. 45. Dordrecht: D. Reidel.

Delsemme, A. H. (ed.). 1977. *Comets, Asteroids, Meteorites: Interrelations, Evolution and Origins*. Toledo: University of Toledo Press.

Donn, B., Mumma, M., Jackson, W., A'Hearn, M., and Harrington, R. (eds.). 1976. *The Study of Comets*. NASA SP-393. Washington, D.C.: GPO.

Gary, G. A. (ed.). 1975. *Comet Kohoutek*. NASA SP-355. Washington, D.C.: GPO.

Kuiper, G. P., and Roemer, E. (eds.). 1972. *Proceedings of the Tucson Comet Conference*. Tucson: University of Arizona Press.

Lyttleton, R. A. 1953. *The Comets and their Origin*. Cambridge: Cambridge University Press.

Marsden, B. G. 1974. Comets. *Ann. Rev. Astron. Astrophys.* *12*:1–21.

Rahe, J., Donn, B., and Wurm, K. 1969. *Atlas of Cometary Forms*. NASA SP-198. Washington, D.C.: GPO.

Richter, N. B. 1963. *The Nature of Comets*. New York: Dover. (Contains a thorough bibliography of pre–1960 comet literature.)

Whipple, F. L., and Huebner, W. F. 1976. Physical Processes in Comets. *Ann. Rev. Astron. Astrophys.* *14*:143–72.

History and early studies of cometary physics

Alfvén, H. 1957. On the Theory of Comet Tails. *Tellus 9:*92–6.

Aristotle. 1952. Meteorology. *Great Books of the Western World 8:*445. Chicago: Encyclopaedia Britannica.

Armitage, A. 1966. *Edmond Halley*. London: Thomas, Nelson.

Biermann, L. 1951. Kometenschweife und Korpuskularstrahlung. *Z. Astrophys.* *29:*274–86.

1953. Physical Processes in Comet Tails and Their Relation to Solar Activity. *Mem. Soc. Sci. Liège, 4th Ser. 13:*251.

1968. *On the Emission of Atomic Hydrogen in Comets*. JILA Report No. 93. Boulder, Colo.: Joint Institute for Laboratory Astrophysics.

Chambers, G. F. 1909. *The Story of Comets*. Oxford: at the Clarendon Press.

Code, A. D., Houck, T. E., and Lillie, C. F. 1972. Ultraviolet Observations of Comets. In *The Scientific Results from the Orbiting Astronomical Observatory (OAO-2)*, ed. A. D. Code. NASA SP-310. Washington, D.C.: GPO.

Delsemme, A. H., and Swings, P. 1952. Hydrates de gaz dans les noyaux cométaires et les grains interstellaires. *Ann. Astrophys. 15:*1–6.

Donn, B., and Urey, H. C. 1956. On the Mechanism of Comet Outbursts and the Chemical Composition of Comets. *Astrophys. J. 123:*339.

Drake, S., and O'Malley, C. D. (trans.). 1960. *The Controversy on the Comet of 1618.* Philadelphia: University of Pennsylvania Press. (Writings by Galileo, Kepler, Griassi, and Guiducci.)

Dreyer, J. L. E. 1953. *A History of Astronomy from Thales to Kepler.* New York: Dover.

FitzGerald, G. F. 1892. *The Electrician 30:*481.

Galileo, 1623. The assayer. In *Discoveries and Opinions of Galileo,* trans. S. Drake. Garden City, N.Y.: Doubleday, 1957.

Harwit, M., and Hoyle, F. 1962. Plasma Dynamics in Comets. II. Influence of Magnetic Fields. *Astrophys. J. 135:*875–82.

Hellman, C. D. 1944. *The Comet of 1577: Its Place in the History of Astronomy.* New York: AMS Press. (A very good reference for the history of cometary thought through the sixteenth century.)

Herschel, J. F. W. 1871. *Outlines of Astronomy.* London: Longmans, Green.

Hertzberg, G., and Lew, H. 1974. Tentative Identification of the H_2O^+ Ion in Comet Kohoutek. *Astron. Astrophys. 31:*123–4.

Hoffmeister, C. 1943. Physikalische Untersuchungen an Kometen. I. Die Beziehungen des primären Schweifstrahls zum Radiusvector. *Z. Astrophys. 22:*265–85.

1944. Physikalische Untersuchungen an Kometen. II. Die Bewegung der Schweif-materie und die Repulsivkraft der Sonne beim Kometen 1942g. *Z. Astrophys. 23:*1–18.

Hoyle, F., and Harwit, M. 1962. Plasma Dynamics in Comets. I. Plasma Instability. *Astrophys. J. 135:*867–74.

Lodge, O. 1900. *The Electrician 46:*249.

Lyttleton, R. A. 1953. *The Comets and their Origin.* Cambridge: Cambridge University Press.

Newton, I. 1686. *Principia.* Trans. by A. Mott. Berkeley: University of California Press, 1962.

Olivier, C. P. 1930. *Comets.* Baltimore: Williams & Wilkins.

Oort, J. H. 1950. The Structure of the Cloud of Comets Surrounding the Solar System and a Hypothesis Concerning Its Origin. *Bull. Astron. Inst. Neth. 11:*91–110.

1951. Origin and Development of Comets. *Observatory 71:*129–44.

Oort, J. H., and Schmidt, M. 1951. Differences Between New and Old Comets. *Bull. Astron. Inst. Neth. 11:*259–70.

Öpik, E. 1932. Note on Stellar Perturbations of Nearly Parabolic Orbits. *Proc. Am. Acad. Arts and Sciences 67:*169–82.

Pannekoek, A. 1961. *A History of Astronomy.* New York: Interscience.

Proctor, M., and Crommelin, A. C. D. 1937. *Comets: Their Nature, Origin and Place in the Science of Astronomy.* London: Technical Press.

Richter, N. B. 1963. *The Nature of Comets.* London: Methuen.

Russell, H. N., Dugan, R. S., and Stuart, J. Q. 1945. *Astronomy. I. The Solar System.* Boston: Ginn.

Schwarzschild, K., and Kron, E. 1911. On the Distribution of Brightness in the Tail of Halley's Comet. *Astrophys. J. 34:*342–52.

Seneca, L. 1910. *Questiones naturales.* Clarke, J. (trans.). London: Macmillan.

Swings, P. 1943. Cometary spectra. *Mon. Not. R. Astron. Soc. 103:*86–112.

1962. Comportement des raies interdites de l'oxygènes dans les comètes. I. Observations. *Ann. Astrophys. 25:* 165–70.

van Woerkom, A. J. J. 1948. On the Origin of Comets. *Bull. Astron. Inst. Neth. 10:*445–72.

Vsekhsvyatskij, S. K. 1977. Comets and the Cosmogony of the Solar System. In *Comets,*

Asteroids, Meteorites: Interrelations, Evolution and Origins, ed. A. H. Delsemme. Toledo: University of Toledo Press.

Whipple, F. L. 1950a. A Comet Model. I. The Acceleration of Comet Encke. *Astrophys. J. 111:*375–94.

   1950b. A Comet Model. II. Physical Relations for Comets and Meteors. *Astrophys. J. 113:*464–74.

   1955. A Comet Model. III. The Zodiacal Light. *Astrophys. J. 121:*750–70.

   1976. Background of Modern Comet Theory. *Nature 263:*15–19.

Zanstra, H. 1929. The Excitation of Line- and Band-spectra in Comets by Sunlight. *Mon. Not. R. Astron. Soc. 89:*178–97.

Discovery, orbits, and dynamics

Brouwer, D., and Clemence, G. M. 1961. *Methods of Celestial Mechanics.* New York: Academic Press.

Delsemme, A. H. (ed.). 1977. *Comets, Asteroids, Meteorites: Interrelations, Evolution and Origins.* Toledo: University of Toledo Press.

Dubyago, A. D. 1961. *The Determination of Orbits.* New York: Macmillan.

Escobal, P. R. 1965. *Methods of Orbit Determination.* New York: John Wiley.

Herget, P. 1948. *The Computation of Orbits.* Cincinnati: University of Cincinnati Press.

Marsden, B. G. 1968. Comets and Nongravitational Forces. I. *Astron. J. 73:*367–79.

   1969. Comets and Nongravitational Forces. II. *Astron. J. 74:*720–34.

   1970. Comets and Nongravitational Forces. III. *Astron. J. 75:*75–84.

   1972. Comets in 1971. *Quart. J. R. Astron. Soc. 13:*415–34.

   1973. Comets in 1972. *Quart. J. R. Astron. Soc. 14:*389–406.

   1974a. Comets. *Ann. Rev. Astron. Astrophys. 12:*1–21. (Contains studies of orbital statistics.)

   1974b. Comets in 1973. *Quart. J. R. Astron. Soc. 15:*433–60.

   1979. *Catalogue of Cometary Orbits.* Cambridge, Mass.: S.A.O. Central Bureau for Astronomical Telegrams.

Marsden, B. G., and Roemer, E. 1978a. Comets in 1974. *Quart. J. R. Astron. Soc. 19:*38–58.

   1978b. Comets in 1975. *Quart. J. R. Astron. Soc. 19:*59–89.

Marsden, B. G., and Sekanina, Z. 1971. Comets and Nongravitational Forces. IV. *Astron. J. 76:*1135–51.

Marsden, B. G., Sekanina, Z., and Yeomans, D. K. 1973. Comets and Nongravitational Forces. V. *Astron. J. 78:*211–25.

Roemer, E. 1963. Comets: discovery, orbits, astrometric observations. In *The Moon, Meteorites, and Comets,* ed. B. M. Middlehurst and G. P. Kuiper. Chicago: University of Chicago Press.

Yeomans, D. K. 1971. Nongravitational Forces Affecting the Motions of Periodic Comets Giacobini–Zinner and Borrelly. *Astron. J. 76:*83–6.

   1974. The Nongravitational Motion of Comet Kopff. *Pub. Astron. Soc. Pacific 86:*125–7.

Photometry and spectroscopy

Arpigny, C. 1965. Spectra of Comets and their Interpretation. *Ann. Rev. Astron. Astrophys. 3:*351–76.

   1977. On the Nature of Comets. *Proceedings of the Robert A. Welch Foundation Conferences on Chemical Research XXI, Cosmochemistry.* Houston.

Biermann, L. 1970. Physical studies of comets. *Trans. IAU. 14A:*141–52.

Bobrovnikoff, N. T. 1941. Investigations of the Brightness of Comets. *Popular Astronomy 49:*467–79.

Brownlee, D. E., Rajan, R. S., and Tomandl, D. A. 1977. A chemical and textural comparison between carbonaceous chondrites and interplanetary dust. In *Comets, Asteroids, Meteorites,* ed. A. H. Delsemme. Toledo: University of Toledo Press.

Brownlee, D. E., Tomandl, D. A., and Hodge, P. W. 1976. Extraterrestrial particles in the stratosphere. In *Interplanetary Dust and Zodiacal Light,* ed. H. Elsasser and H. Fechtig. New York: Springer-Verlag.

Donn, B., Mumma, M., Jackson, W., A'Hearn, M., and Harrington, R. (eds.). 1976. *The Study of Comets.* NASA SP-393. Washington, D.C.: GPO.

Greenstein, J. L. 1958. High-Resolution Spectra of Comet Mrkos (1957d). *Astrophys. J. 128:*106–13. (This paper originates the Greenstein effect.)

Kuiper, G. P., and Roemer, E. (eds.). 1972. *Proceedings of the Tucson Comet Conference.* Tucson: University of Arizona Press.

Lanzerotti, L. J., Robbins, M. F., Tolk, N. H., and Neff, S. H. 1974. High Resolution Scans of Comet Kohoutek in the Vicinity of 5015, 5890 and 6563 Å. *Icarus 23:*618–22.

Maas, R. W., Ney, E. P., and Woolf, N. J. 1970. The 10-micron Emission Peak of Comet Bennett 1969i. *Astrophys. J. 160:*L101–4.

Marsden, B. G. 1974. Comets. *Ann. Rev. Astron. Astrophys. 12:*1–21.

Ney, E. P. 1974. Multiband Photometry of Comets Kohoutek, Bennett, Bradfield and Encke. *Icarus 23:*551–60.

Noguchi, K., Sato, S., Maihara, T., Okuda, H., and Uyama, K. 1974. Infrared Photometric and Polarimetric Observations of Comet Kohoutek. *Icarus 23:*545–50.

Parcel, L. G., and Beavers, W. I. 1974. Emission Band Ratios in Comet Kohoutek (1973f). *Icarus 23:*623–9.

Rahe, J. 1980. Ultraviolet spectroscopy of comets. In *Second European IUE Conference,* ed. B. Battrick and J. Mort. ESA SP-157.

Swings, P. 1943. Cometary Spectra. *Mon. Not. R. Astron. Soc. 103:*86–112.

1965. Cometary Spectra. *Quart. J. R. Astron. Soc. 6:*28–69.

Traub, W. A., and Carleton, N. P. 1974. A Search for H_2O and CH_4 in Comet Kohoutek. *Icarus 23:*585–9.

Vsekhsvyatskij, S. K. 1958, 1966, 1967. *Fizicheskie Karakteristiki Komet.* Moscow: Nauka. (Three Russian volumes describing physical characteristics of comets.)

Whipple, F. L., and Huebner, W. F. 1976. Physical Processes in Comets. *Ann. Rev. Astron. Astrophys. 14:*143–72.

Zeilik, M. II, and Wright, E. L. 1974. Infrared Photometry of Comet Kohoutek. *Icarus 23:*577–9.

Structure of comets

Arpigny, C. 1977. On the Nature of Comets. *Proceedings of the Robert A. Welch Foundation Conferences on Chemical Research XXI, Cosmochemistry.* Houston.

Belton, M. J. S. 1965. Some Characteristics of Type II Comet Tails and the Problem of Distant Comets. *Astron. J. 70:*451–65.

Bertaux, J. L., Blamont, J. E., and Festou, M. 1973. Interpretation of Hydrogen Lyman-Alpha Observations of Comets Bennett and Encke. *Astron. Astrophys. 25:*415–30.

Biermann, L. 1951. Kometenschweife und Korpuskularstrahlung. *Z. Astrophys.* *29:*274–86. (On comet tails and the solar wind.)

1968. On the Emission of Atomic Hydrogen in Comets. JILA Report No. 93. Boulder, Colo.: Joint Institute for Laboratory Astrophysics.

Biermann, L., Brosowski, B., and Schmidt, H. U. 1967. The Interaction of the Solar Wind with a Comet. *Sol. Phys. 1:*254–84.

Brandt, J. C. 1968. The Physics of Comet Tails. *Ann. Rev. Astron. Astrophys.* *6:*267–86.

Brandt, J. C., and Mendis, D. A. 1979. The interaction of the solar wind with comets. In *Solar System Plasma Physics,* Vol. II, ed. C. F. Kennel, L. J. Lanzerotti, and E. N. Parker. Amsterdam: North-Holland.

Brandt, J. C., Roosen, R. G., and Harrington, R. S. 1972. An Astrometric Determination of Solar Wind Velocities from Orientations of Ionic Comet Tails. *Astrophys. J.* *177:*277–84.

Carruthers, G. R., Opal, C. B., Page, T. L., Meier, R. R., and Prinz, D. K. 1974. Lyman-α Imagery of Comet Kohoutek. *Icarus 23:*526–37.

Chebotarev, G. A., Kazimirchaka-Polonskaya, E. I., and Marsden, B. G. (eds.). 1972. *The Motion, Evolution of Orbits and Origins of Comets.* IAU Symposium No. 45. Dordrecht: D. Reidel.

Code, A. D., Houck, T. E., and Lillie, C. F. 1972. Ultraviolet observations of comets. In *The Scientific Results from the Orbiting Astronomical Observatory (OAO-2),* ed. A. D. Code. NASA SP-310. Washington, D.C.: GPO. (Observations of the hydrogen cloud.)

Combi, M. R., and Delsemme, A. H. 1980a. Neutral Cometary Atmospheres. I. An Average Random Walk Model for Photodissociation in Comets. *Astrophys. J.* *237:*663–40.

1980b. Neutral Cometary Atmospheres. II. The Production of CN in Comets. *Astrophys. J. 237:*641–5.

Delsemme, A. H. 1966. Vers un modèle physico-chimique du noyau cométaire. In *Nature et origine des comètes. Mém. Soc. R. Sci. Liège 12:*77–110.

1973a. Gas and dust in comets. In *Cosmochemistry,* ed. A. G. W. Cameron. Dordrecht: D. Reidel.

1973b. Brightness Law of Comets. *Astrophys. Lett. 14:*163–7.

1975. The Volatile Fraction of the Cometary Nucleus. *Icarus 24:*95–110.

Delsemme, A. H., and Delsemme, J. 1971. Private communication on vaporization rates. (See Marsden, B. G., Sekanina, Z., and Yeomans, D. K. 1973. *Astron. J. 78:*211–25 for details.)

Delsemme, A. H., and Miller, D. C. 1970. Physico-chemical Phenomena in Comets – II. Gas Adsorption in the Snows of the Nucleus. *Planet. Space Sci. 18:*717–30.

1971a. Physico–Chemical Phenomena in Comets – III. The Continuum of Comet Burnham (1960 II). *Planet. Space Sci. 19:*1229–57.

1971b. Physico–Chemical Phenomena in Comets – IV. The C_2 Emission of Comet Burnham (1960 II). *Planet. Space Sci. 19:*1259–74.

Delsemme, A. H., and Rud, D. A. 1973. Albedos and Cross Sections for Nuclei of Comets 1969 IX, 1970 II and 1971 I. *Astron. Astrophys. 28:*1–6.

Delsemme, A. H., and Swings, P. 1952. Hydrates de gaz dans les noyaux cométaires et les grains interstellaires. *Ann. Astrophys. 15:*1–6.

Delsemme, A. H., and Wenger, A. 1970. Physico-chemical Phenomena in Comets – I. Experimental Studies of Snows in a Cometary Environment. *Planet. Space Sci. 18:*709–15.

Donn, B., Mumma, M., Jackson, W., A'Hearn, M., and Harrington, R. (eds.). 1976. *The*

Study of Comets. NASA SP-393. Washington, D.C.: GPO. (Contains several articles on comas of comets.)

Festou, M. 1978. L'hydrogène atomique et la radical oxhydrille dans les comètes. Thèse de doctorat d'Etat, université Pierre et Marie Curie, Paris.

Finson, M. L., and Probstein, R. F. 1968a. A Theory of Dust Comets. I. Model and Equations. *Astrophys. J. 154:*327–52.

1968b. A Theory of Dust Comets. II. Results for Comet Arend–Roland. *Astrophys. J. 154:*353–80.

Gary, G. A., and O'Dell, C. R. 1974. Interpretation of the Anti-Tail of Kohoutek as a Particle Flow Phenomenon. *Icarus 23:*519–25.

Haser, L. 1957. Distribution d'intensité dans la tête d'une comète. *Bull. Acad. R. Belg., 5ᵉ série 43:*740.

Huebner, W. F. 1970. Dust from Cometary Nuclei. *Astron. Astrophys. 5:*286–97.

Hughes, D. W. 1975. Cometary Outbursts, a Brief Survey. *Quart. J. R. Astron. Soc. 16:*410–27.

Hyder, C. L., Brandt, J. C., and Roosen, R. G. 1974. Tail Structures Far from the Head of Comet Kohoutek. *Icarus 23:*601–10.

Ip, W.-H., and Mendis, D. A. 1974. Neutral Atmospheres of Comets: A Distributed Source Model. *Astrophys. Space Sci. 26:*153–66.

Jackson, W. M., and Donn, B. D. 1966. Collisional Processes in the Inner Coma. In *Nature et origine des comètes. Mém. Soc. R. Sci. Liège 12:*133–40.

Keller, H. U. 1973. Hydrogen Production Rates of Comet Bennett (1969i) in the First Half of April 1970. *Astron. Astrophys. 27:*51–7.

Liller, W. 1960. The Nature of Grains in the Tails of Comets 1956h and 1957d. *Astrophys. J. 132:*867–82.

Malaise, D. J. 1970. Collisional Effects in Cometary Atmospheres I. Model Atmospheres and Synthetic Spectra. *Astron. Astrophys. 5:*209–27.

Osterbrock, D. E. 1958. A Study of Two Comet Tails. *Astrophys. J. 128:*95–105.

Potter, A. E., and Del Duca, B. 1964. Lifetime in Space of Possible Parent Molecules of Cometary Radicals. *Icarus 3:*103–8.

Probstein, R. N. 1968. The Dusty Gasdynamics of Comet Heads. In *Problems of Hydrodynamics and Continuum Mechanics,* ed. Society for Industrial Mathematics, Philadelphia, p. 578.

Roemer, E. 1963. Comets: Discovery, Orbits, Astrometric Observations. In *The Moon, Meteorites and Comets,* ed. B. M. Middlehurst and G. P. Kuiper. Chicago: University of Chicago Press.

Sekanina, Z. 1974. On the Nature of the Anti-Tail of Comet Kohoutek (1973f). I. A Working Model. *Icarus 23:*502–18.

1975. A Study of the Icy Tails of Distant Comets. *Icarus 25:*218–38.

1977. Relative Motions of Fragments of Split Comets. I. *Icarus 30:*574–94.

1978. Relative Motions of Fragments of Split Comets. II. *Icarus 33:*173–85.

Sekanina, Z., and Miller, F. D. 1973. Comet Bennett 1970 II. *Science 179:*565–7.

Stefanik, R. P. 1966. On Thirteen Split Comets. In *Nature et Origine Des Comètes. Mém. Soc. R. Sci. Liège 12:*29–32.

Whipple, F. L. 1955. A Comet Model. III. The Zodiacal Light. *Astrophys. J. 121:*750–70.

1957. Comments on the Sunward Tail of Comet Arend–Roland. *Sky and Telescope. 16:*246–8.

1961. Problems of the Cometary Nucleus. *Astron. J. 66:*375–80.

Whipple, F. L., and Huebner, W. F. 1976. Physical Processes in Comets. *Ann. Rev. Astron. Astrophys. 14:*143–72.

Models and origins of comets

Biermann, L., and Lüst, R. 1972. Some New Results on the Plasma–Dynamical Processes near Comets. Max-Planck-Institut für Physik und Astrophysik Report MPI/PAE-Astro 52.

Biermann, L., Brosowski, B., and Schmidt, H. U. 1967. The Interaction of the Solar Wind with a Comet. *Sol. Phys. 1:*254–84.

Brosowski, B., and Wegmann, R. 1972. Numerische Behandlung eines Kometmodells. Max-Planck-Institut für Physik und Astrophysik Report MPI/PAE-Astro 46.

Cameron, A. G. W. 1973. Accumulation Processes in the Primitive Solar Nebula. *Icarus 18:*407–50.

Cameron, A. G. W., and Pine, M. R. 1973. Numerical Models of the Primitive Solar Nebula. *Icarus 18:*377–406.

Chebotarev, G. A., Kazimirchaka-Polonskaya, E. I., and Marsden, B. G. (eds.). 1972. *The Motion, Evolution of Orbits and Origins of Comets.* IAU Symposium No. 45. Dordrecht: D. Reidel.

Cherednichenko, V. I. 1974. On Some New Sources of Formation of Molecules Observed in Cometary Atmospheres. *Problems of Cosmic Physics 9:*155–58. (In Russian.)

Code, A. D., Houck, T. E., and Lillie, C. F. 1972. Ultraviolet Observations of Comets. In *The Scientific Results from the Orbiting Astronomical Observatory (OAO-2)*, ed. A. D. Code. NASA SP-310. Washington, D.C.: GPO.

Delsemme, A. H. 1973a. Origin of the Short-Period Comets. *Astron. Astrophys. 29:*377–81.

1973b. Brightness Law of Comets. *Astrophys. Lett. 14:*163–7.

1975. The Volatile Fraction of the Cometary Nucleus. *Icarus 24:*95–110.

(ed.). 1977a. *Coments, Asteroids, Meteorites: Interrelations, Evolution and Origins.* Toledo: University of Toledo Press.

1977b. The pristine nature of comets. In *Comets, Asteroids, Meteorites: Interrelations, Evolution and Origins*, ed. A. H. Delsemme. Toledo: University of Toledo Press.

Delsemme, A. H., and Rud, D. A. 1973. Albedos and Cross Sections for Nuclei of Comets 1969 IX, 1970 II and 1971 I. *Astron. Astrophys. 28:*1–6.

Delsemme, A. H., and Swings, P. 1952. Hydrates de gaz dans les noyaux cométaires et les grains interstellaires. *Ann. Astrophys. 15:*1–6.

Delsemme, A. H., and Wenger, A. 1970. Physico-Chemical Phenomena in Comets – I. Experimental Studies of Snows in a Cometary Environment. *Planet. Space Sci. 18:*709–15.

Donn, B., Mumma, M., Jackson, W., A'Hearn, M., and Harrington, R. (eds.). 1976. *The Study of Comets.* NASA SP-393. Washington, D.C.: GPO.

Everhart, E. 1972. Origin of Short-Period Comets. *Astrophys. Lett. 10:*131–5.

Joss, P. C. 1973. On the Origin of Short-Period Comets. *Astron. Astrophys. 25:*271–73.

Lewis, J. S. 1972a. Metal/Silicate Fractionation in the Solar System. *Earth and Planetary Science Letters 15:*286–90.

1972b. Low Temperature Condensation from the Solar Nebula. *Icarus 16:*241–52.

Oort, J. H. 1950. The Structure of the Cloud of Comets Surrounding the Solar System and a Hypothesis Concerning Its Origin. *Bull. Astron. Inst. Neth. 11:*91–110.

Oppenheimer, M. 1975. Gas Phase Chemistry in Comets. *Astrophys. J. 196:*251–59.

Ovenden, M. W. 1975. Bode's Law – Truth or Consequences. *Vistas in Astronomy 18:*473–96. (Recent thoughts on cometary origins.)

Potter, A. E., and Del Duca, B. 1964. Lifetime in Space of Possible Parent Molecules of Cometary Radicals. *Icarus 3:*103–8.

Report of the Comet Science Working Group, Executive Summary. August 1979. NASA TM 80542.

Sekanina, Z., and Miller, F. D. 1973. Comet Bennett 1970 II. *Science 179:*565–7.

Van Flandern, T. C. 1977. A former major planet of the solar system. In Delsemme (1977).

1978. A Former Asteroidal Planet as the Origin of Comets. *Icarus 36:*51–74. (Van Flandern's papers are recent speculations bearing on the origin of comets.)

van Woerkom, A. J. J. 1948. On the Origin of Comets. *Bull. Astron. Inst. Neth. 10:*445–72.

Whipple, F. L. 1950a. A Comet Model. I. The Acceleration of Comet Encke. *Astrophys. J. 111:*375–94.

1950b. A Comet Model. II. Physical Relations for Comets and Meteorites. *Astrophys. J. 113:*464–74.

1955. A Comet Model. III. The Zodiacal Light. *Astrophys. J. 121:*750–70.

Comets and the solar system

Biermann, L. 1951. Kometenschweife und Korpuskularstrahlung. *Z. Astrophys. 29:*274–86.

Brandt, J. C., and Mendis, D. A. 1979. The interaction of the solar wind with comets. In *Solar System Plasma Physics,* Vol. II, ed. C. F. Kennel, L. J. Lanzerotti, and E. N. Parker. Amsterdam: North-Holland.

Cook, A. F., Halliday, I., and Millman, P. M. 1971. Photometric Analysis of Spectrograms of Two Perseid Meteors. *Can. J. Phys. 49:*1738–49.

Geise, R. H., and Grün, E. 1976. The compatibility of recent micrometeoroid flux curves with observations and models of the zodiacal light. In *Interplanetary Dust and Zodiacal Light,* ed. H. Elsässer and H. Fechtig. Berlin: Springer-Verlag.

Goldberg, R. A., and Aikin, A. C. 1973. Comet Encke: Meteor Metallic Ion Identification by Mass Spectrometry. *Science 180:*294.

Hoffmeister, C. 1943. Physikalische Untersuchungen an Kometen. I. Die Beziehungen des primären Schweifstrahls zum Radiusvector. *Z. Astrophys. 22:*265–85.

1944. Physikalische Untersuchung an Kometen. II. Die Bewegung der Schweifmaterie und die Repulsivkraft der Sonne biem Kometen 1942g. *Z. Astrophys. 23:*1–18.

Miller, F. D. 1976. Solar–Cometary Relations and the Events of June–August 1972. *Space Sci. Rev. 19:*739–59.

Millman, P. 1971. Cometary Meteroids. *Nobel Symposium No. 21:*157–66.

Whipple, F. L. 1955. A Comet Model. III. The Zodiacal Light. *Astrophys. J. 121:*750–70.

Recent developments

Ananthakrishnan, S., Bhandari, S. M., and Rao, A. P. 1975. Occultation of Radio Source PKS 2025-15 by Comet Kohoutek (1973f). *Astrophys. Space Sci. 37:*275–82.

Astronomy and Astrophysics Abstracts (New York: Springer-Verlag) is an abstracting periodical that has been published since 1969. The periodical has a section devoted to comets in each semiannual issue. The reader is referred to that publication for the most recent literature on comets.

Brandt, J. C., Hawley, J. D., and Niedner, M. B., Jr. 1980. A Very Rapid Turning of the Plasma-Tail Axis of Comet Bradfield 1979I on February 6, 1980. *Astrophys. J. Lett. 241:*L51–4.

Brandt, J. C., and Mendis, D. A. 1979. The interaction of the solar wind with comets. In *Solar System Plasma Physics,* Vol. II, ed. C. F. Kennel, L. J. Lanzerotti, and E. N. Parker. Amsterdam: North-Holland.

Bruston, P., Coron, N., Dambier, G., Laurent, C., Leblanc, J., Lena, P., Rather, J. D. G., and Vidal-Madjar, A. 1974. Observations of Comet Kohoutek at 1.4 mm. *Nature 252:*665–6.

Cherednichenko, V. I. 1974. On Some New Sources of Formation of Molecules Observed in Cometary Atmospheres. *Problems of Cosmic Physics 9:*155–58. (In Russian.)

Cosmovici, C. B., Strafella, F., Kimagli, L., D'Innocenzo, A., Laggieri, G., Nesti, C., and Perrone, A. 1978. Splitting of Comet West 1975n: Photometry and Narrow-Band Photometry. *Astron. Astrophys. 63:*83–6.

Delsemme, A. H. 1976. Chemical Nature of the Cometary Snows. *Mém. Soc. R. Sci. Liège, 6ᵉ série 9:*135–45.

(ed.). 1977. *Comets, Asteroids, Meteorites: Interrelations, Evolution and Origins.* Toledo: University of Toledo Press.

Donn, B., Mumma, M., Jackson, W., A'Hearn, M., and Harrington, R., (eds.). 1976. *The Study of Comets.* NASA SP-393. Washington, D.C.: GPO.

Ershkovich, A. I. 1978. The Comet Tail Magnetic Field: Large or Small? *Mon. Nat. R. Astron. Soc. 184:*755–8.

Farquhar, R. W. 1974. Mission Strategy for Cometary Exploration in the 1980's. NASA: Goddard Space Flight Center, Document X-581-74-337.

Hobbs, R. W., Maran, S. P., Brandt, J. C., Webster, W. J., and Krishna-Swamy, K. S. 1975. Microwave Continuum Radiation from Comet Kohoutek 1973f: Emission from the Icy Grain Halo. *Astrophys. J. 201:*749–55.

Hyder, C. L., Brandt, J. C., and Roosen, R. G. 1974. Tail Structures far from the Head of Comet Kohoutek. I. *Icarus 23:*601–10.

Ip, W.-H., and Mendis, D. A. 1975. The Cometary Magnetic Field and Its Associated Electric Currents. *Icarus 26:*457–61.

1976. The Generation of Magnetic Fields and Electric Currents in Cometary Plasma Tails. *Icarus 29:*147–51.

1978. The Flute Instability as the Trigger Mechanism for Disruption of Cometary Plasma Tails. *Astrophys. J. 223:*671–5.

Keller, H. U., and Lillie, C. F. 1978. Hydrogen and Hydroxyl Production Rates of Comet Tago–Sato–Kosaka. *Astron. Astrophys. 62:*143–7.

Maran, S. P., and Hobbs, R. W. (eds.). 1974. *Special Issue on Comet Kohoutek. Icarus 23:*No. 4 (December).

Mendis, D. A., and Wickramasinghe, N. C. 1975. Composition of Cometary Dust: The Case Against Silicates. *Astrophys. Space Sci. 37:*L13–16.

Neugebauer, M., Yeomans, D. K., Brandt, J. C., and Hobbs, R. W. 1977. *Space Missions to Comets.* NASA CP-2089. Washington, D.C.: GPO.

Niedner, M. B., and Brandt, J. C. 1978. Interplanetary Gas. XXIII. Plasma Tail Disconnection Events in Comets: Evidence for Magnetic Field Line Reconnection at Interplanetary Sector Boundaries? *Astrophys. J. 223:*655–70.

Niedner, M. B., Rothe, E. D., and Brandt, J. D. 1978. Interplanetary Gas. XXII. Interaction of Comet Kohoutek's Ion Tail with the Compression Region of a Solar-Wind Co-rotating Stream. *Astrophys. J. 221:*1014–25.

Opal, C. B., and Carruthers, G. R. 1978. Lyman-α Observations of Comet West. *Icarus 31:*503–9.

Oppenheimer, M. 1975. Gas Phase Chemistry in Comets. *Astrophys. J. 196:*251–59.

Sekanina, Z. 1977. Relative Motions of the Fragments of Split Comets. I. *Icarus 30:*574–94.

1978. Relative motions of the Fragments of Split Comets. II. *Icarus 33:*173–85.
Sekanina, Z., and Pansecchi, L. 1977. The Anti-Tail of Comet Bradfield. *Astrophys. Lett. 18:*61–3.

Comets and man

The literature for this part is somewhat skimpy. Some of the sources are of general interest.

Alter, D. 1956. Comets and People. *Griffith Observer 20:*74–82.
Baxter, J., and Atkins, T. 1976. *The Fire Came By.* Garden City, N.Y.: Doubleday.
Brown, J. C., and Hughes, D. W. 1977. Tunguska's Comet and Non-thermal ^{14}C Production in the Atmosphere. *Nature 268:*512–14.
Brown, P. L. 1974. *Comets, Meteorites and Man.* New York: Taplinger.
Chambers, G. F. 1909. *The Story of Comets.* Oxford: Clarendon Press.
Chang, S. 1979. Comets: cosmic connections with carbonaceous meteorites, interstellar molecules and the origins of life. In *Space Missions to Comets,* ed. M. Neugebauer, D. K. Yeomans, J. C. Brandt, and R. W. Hobbs. NASA CP-2089. Washington, D.C.: GPO.
Durant, W. 1954. *The Story of Civilization I. Our Oriental Heritage.* New York: Simon & Schuster.
Fesenkov, V. G. 1966. A Study of the Tunguska Meteorite Fall. *Soviet Astronomy – A. J. 10:*195–213.
French, B. M. 1978. Comet or Spacecraft? *Secondlook,* December, pp. 1–3. (On the Tunguska event.)
Goodavage, J. F. 1973. *The Comet Kohoutek.* New York: Pinnacle Books.
Haney, H. G. 1965. Comets: A Chapter in Science and Superstition in Three Golden Ages: The Aristotelian, the Newtonian, and the Thermonuclear. Thesis, University of Alabama. Ann Arbor, Mich.: University Microfilms.
Hellman, C. D. 1944. *The Comet of 1577: Its Place in the History of Astronomy.* New York: AMS Press.
Hoyle, F., and Wickramasinghe, N. C. 1978a. *Lifecloud: The Origin of Life in the Universe.* New York: Harper & Row. (Discusses comets and their connection with the origin of life.)
    1978b. Influenza from Space. *New Scientist 79:*946–8.
Krinov, E. L. 1963. The Tunguska and Sikhote–Alin meteorites. In *The Moon, Meteorites and Comets,* ed. B. M. Middlehurst and G. P. Kuiper. Chicago: University of Chicago Press.
Marsden, B. G. 1977. Comet Halley and history. In *Space Missions to Comets,* ed. M. Neugebauer, D. K. Yeomans, J. C. Brandt, and R. W. Hobbs. NASA CP-2089. Washington, D.C.: GPO.
Niven, L., and Pournelle, J. 1977. *Lucifer's Hammer.* New York: Fawcett-Crest.
Olson, R. J. M. 1979. Giotto's Portrait of Halley's Comet. *Scientific American 240:*160–170. (Discusses art and comets.)
Ridpath, I. 1977a. The Comet that Hit the Earth. *Mercury 6:*2–5.
    1977b. Tunguska the Final Answer. *New Scientist 75:*346–7.

Index of comets

Index